Chapter 1
Background and Motivation

1.1 Magnetosphere

Magnetosphere is the vast plasma region controlled by the Earth's magnetic field. The boundary separating the shocked solar wind from the magnetospheric plasma is the magnetopause. The long trailing limb of the magnetosphere swept by the solar wind away from the sun is the magnetotail. A large number of charged particles with energies from $\sim$eV to 10^8 eV are trapped in the inner magnetosphere. These trapped particles exhibit significant variability in response to the solar wind disturbances and/or due to their spontaneous instabilities. According to the energy, these trapped particles are classified into the plasmasphere ($\sim$eV), ring current ($\sim$1–100 keV), and Van Allen radiation belt ($\sim >$100 keV).

1.2 Electron Radiation Belt Dynamics

The Van Allen radiation belt was discovered by Allen and Frank (1959). The radiation belt can be divided into two zones centering $L \approx 1.8$ and 4.5 separated by the slot region (Lyons et al. 1972; Lyons and Thorne 1973). The inner radiation belt mainly comprises the energetic protons and appears to be relatively stable. In contrast, the outer radiation belt primarily consists of energetic electrons and often exhibits the complex and dramatic dynamics (e.g., Paulikas and Blake 1979; Baker et al. 1986). These energetic particles can significantly harm the space-borne systems through the deep-dielectric charging, the surface charging, and the single-event upsets (Baker et al. 1994; Baker 2002).

Z. Su, *A Global Kinetic Model for Electron Radiation Belt Formation and Evolution*, Springer Theses, DOI 10.1007/978-3-662-46651-3_1

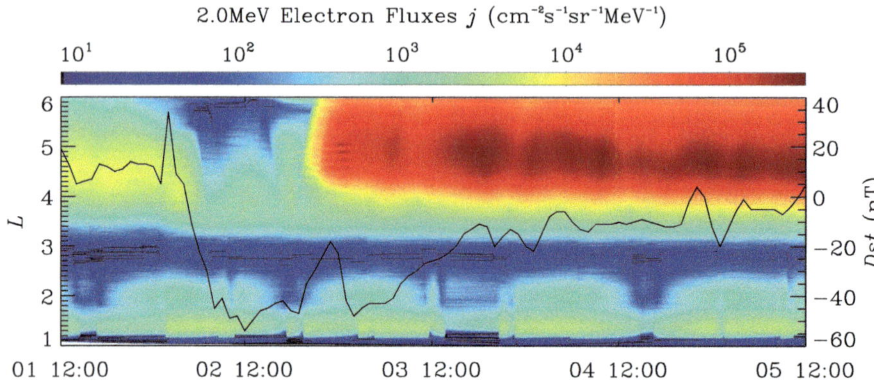

Fig. 1.1 Evolution of 2.0 MeV electron fluxes observed by the Van Allen Probes during the 2 October 2013 storm with the *black line* for the *Dst* index

1.2.1 Response to Geomagnetic Storms

The typical evolution characteristics of outer radiation belt (see Fig. 1.1) during geomagnetic storms include the rapid dropout during the main phase and the gradual buildup during the recovery phase (e.g., Baker et al. 1989; Kim and Chan 1997; Brautigam and Albert 2000; Friedel et al. 2002; Bortnik et al. 2006; Albert et al. 2009; Su et al. 2011). Reeves et al. (2003) statistically analyzed the response of outer radiation belt electrons to the geomagnetic storms. It was found that, 53 % of storms increased the fluxes of relativistic electrons, 19 % of storms decreased the fluxes, and 28 % of storms produced little or no change in the fluxes. These statistical results suggest that, there simultaneously exist various acceleration and loss mechanisms, and the competition of these mechanisms determines the final variation in energetic electron fluxes. The dominance of acceleration mechanism leads to the net increase; the dominance of loss mechanism results in the net decrease; the equivalence of loss and acceleration mechanisms changes little.

1.2.2 Response to Substorms

During the substorms, the electrons can be transported from the magnetotail to the inner magnetosphere, causing the simultaneous enhancement of radiation belt electron fluxes over a wide energy range within several minutes (e.g., McIlwain 1974; Mauk and McIlwain 1974; Konradi et al. 1975; Mauk and Meng 1983; Reeves et al. 1996). Figure 1.2 shows an example of substorm injection observed by the LANL 1995-095 satellite at the geostationary orbit on 10 January 1997. Clearly, within about three minutes, the electron fluxes increased by about two orders of magnitude at the energies 50–75 keV and by about one order of magnitude at the energies ~100 keV.

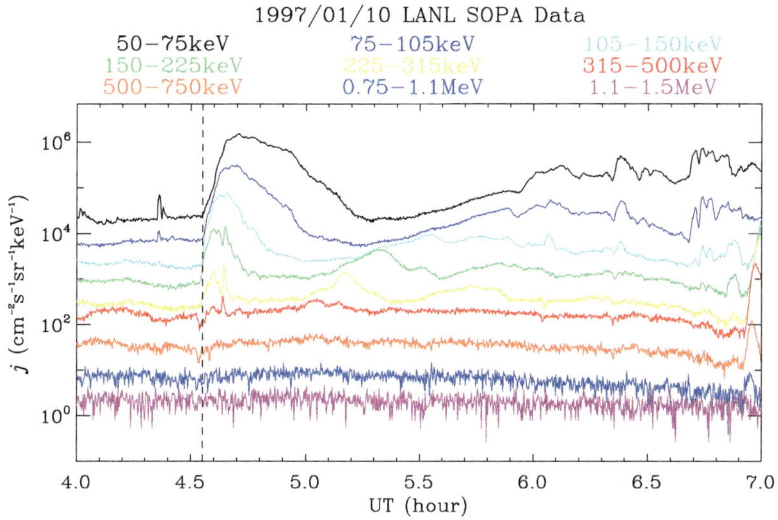

Fig. 1.2 Evolution of energetic electron fluxes observed by the LANL 1995-095 satellite at the geostationary orbit on 10 January 1997

1.2.3 Response to Solar Cycles and Seasons

Li et al. (2001) provided a long-term global picture of the radiation belts based on the SAMPEX observations. As shown in Fig. 1.3, the outer radiation belt electrons show a strong seasonal and solar cycle variation. On a timescale of solar cycle, the energetic electron fluxes maximize during the descending/ascending phase, and minimize during the solar minimum. On a timescale of season, the energetic electron fluxes peak around the equinoxes.

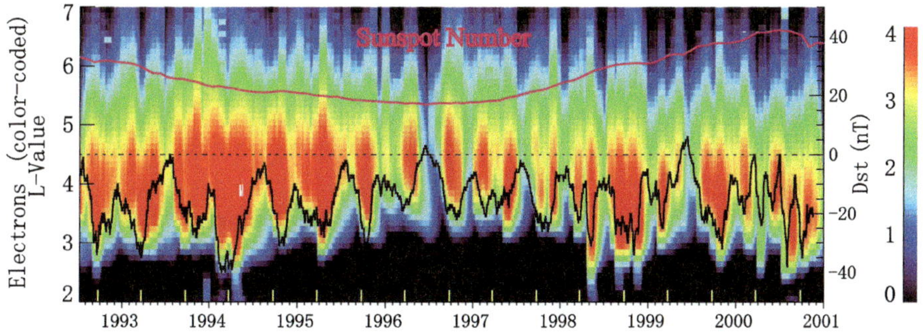

Fig. 1.3 Monthly-averaged 2–6 MeV electron fluxes, the *Dst* index and the sunspot number in the time range from July 1992 to December 2000. Reprinted with permission from (Li et al. 2001). Copyright [2001] [John Wiley and Sons]

1.3 Basic Theory

1.3.1 Single-Particle Orbit Theory

The basic equation describing the motions of charged particles in the electromagnetic fields is written as

$$\frac{\mathrm{d}\mathbf{p}}{\mathrm{d}t} = q\,(\mathbf{E} + \mathbf{v} \times \mathbf{B}), \tag{1.1}$$

where $\mathbf{p}$ is the particle relativistic momentum, q is the particle charge, $\mathbf{v}$ is the velocity, $\mathbf{E}$, and $\mathbf{B}$ are the background electromagnetic fields. In most cases, there is no analytical solution to this differential equation. The motions of trapped particles in the inner magnetosphere can be approximately decomposed into three periodic motions: the gyration around field lines, bounce along the magnetic field lines, and drift circling the Earth.

In the classical Hamiltonian theory, there is a set of canonical coordinates, action variables $\mathbf{J} = (J_1,\ J_2,\ J_3)$, and angle variables $\boldsymbol{\varphi} = (\varphi_1,\ \varphi_2,\ \varphi_3)$, corresponding to the three periodic motions (Schulz and Lanzerotti 1974; Walt 1994; Bourdarie et al. 1997). The three action variables are defined as

$$J_i = \frac{1}{2\pi} \oint_i (\mathbf{p} + q\mathbf{A}) \cdot \mathrm{d}\mathbf{l}, \tag{1.2}$$

where the integration intervals are the periodic motion paths, and $\mathbf{A}$ is the vector potential of magnetic field. Based on the three action variables, we can define three adiabatic invariants, which are approximately constant in the slow-varying background field (Alfven 1950; Northrop 1963; Spitzer 1965).

The first periodic motion is the gyration around the magnetic field lines, whose radius r_g and period τ_g are

$$r_g = \frac{p \sin\alpha}{|q|B}, \tag{1.3}$$

$$\tau_g = \frac{2\pi\gamma m_0}{|q|B}, \tag{1.4}$$

with the Lorentz factor γ, the rest mass m_0, and the pitch-angle α. The first action variable can be obtained as

$$J_1 = \frac{1}{2\pi} \oint (\mathbf{p} + q\mathbf{A}) \cdot \mathrm{d}\mathbf{l} = \frac{p^2 \sin^2\alpha}{2qB}. \tag{1.5}$$

The corresponding adiabatic invariant is written as

$$\mu = \frac{p^2 \sin^2 \alpha}{2m_0 B}, \tag{1.6}$$

which represents the magnetic dipole moment of the current loop produced by the gyration of charged particles in the nonrelativistic limit.

The center of gyration is called the guiding center (Alfven 1950; Spitzer 1965; Northrop 1963). Northrop (1963) had given the motion equation of guiding center as

$$\frac{\mathrm{d}\mathbf{r}}{\mathrm{d}t} = \frac{p_\parallel}{\gamma m_0}\mathbf{b} + \frac{\mu}{q\gamma}\frac{\mathbf{b} \times \nabla B}{B} + \frac{p_\parallel^2}{q\gamma m_0}\frac{\mathbf{b} \times (\mathbf{b} \cdot \nabla)\,\mathbf{b}}{B} + \frac{\mathbf{E} \times \mathbf{b}}{B}, \tag{1.7a}$$

$$\frac{\mathrm{d}p_\parallel}{\mathrm{d}t} = \mathbf{b} \cdot \left(q\mathbf{E} - \frac{\mu}{\gamma}\nabla B\right), \tag{1.7b}$$

where $p_\parallel = p\cos\alpha$ is the parallel momentum with respect to the magnetic field, and $\mathbf{b} = \mathbf{B}/B$ the unit vector along the magnetic field. On the right-hand side of Eq. (1.7a), the first term describes the motion along the magnetic field line, and the other three terms characterize the motions (magnetic gradient drift, magnetic curvature drift, and electric drift) perpendicular to the magnetic field.

The second periodic motion is the bounce motion along the magnetic field lines. As shown in Eq. (1.7b), the particle is exerted a force parallel to the magnetic field

$$\mathbf{F} = -\mu/\gamma \nabla_\parallel B. \tag{1.8}$$

Generally speaking, the magnetic field intensity minimizes at the equator and maximizes at the polar region. Such force F can cause the bounce motion of particles along the magnetic field line. Considering the conservation of first adiabatic invariant in the course of bounce, we can obtain the following relation:

$$\frac{\sin^2 \alpha(s)}{B(s)} = \frac{\sin^2 \alpha_i}{B_i}, \tag{1.9}$$

with the initial pitch-angle α_i and magnetic field B_i, and the pitch-angle $\alpha(s)$ and magnetic field $B(s)$ at a position s along magnetic field line. The magnetic field at the reflection point (mirror point) is

$$B_m = \frac{B_i}{\sin^2 \alpha_i}. \tag{1.10}$$

If the mirror point is in the atmosphere, the charged particles will precipitate due to interaction with the neutral molecules and atoms. In the typical geomagnetic dipole field, the latitude of mirror point increases as the equatorial pitch-angle α_e increases at a fixed magnetic shell L. The particle with the critical equatorial pitch-angle

$$\alpha_L = \arcsin\{L^{-3/2}(4 - 3/L)^{-1/4}\} \tag{1.11}$$

has the mirror point at the surface of the Earth. The particles with smaller pitch-angle would precipitate in the atmosphere. Hence, this angle α_L is named as the loss cone angle. The bounce-period can be calculated as

$$\tau_b = 2\int_{s'_m}^{s_m} \frac{ds}{v\cos\alpha} = \frac{2}{v}\int_{s'_m}^{s_m} \frac{ds}{\left(1 - \frac{B(s)}{B_m}\right)^{\frac{1}{2}}}, \tag{1.12}$$

with the mirror points s_m and s'_m. In the dipole field, it can be approximately expressed as (Hamlin et al. 1961; Roederer 1970)

$$\tau_b = \frac{4LR_E}{\beta c}T(\alpha_e), \tag{1.13}$$

where R_E is the radius of the Earth, $\beta = v/c$, and $T(\alpha_e)$ is the normalized bounce-period

$$T(\alpha_e) \approx 1.30 - 0.56\sin\alpha_e. \tag{1.14}$$

The second action variable is obtained as

$$J_2 = \frac{1}{2\pi}\oint(\mathbf{p} + q\mathbf{A})\cdot \mathbf{ds} = \frac{1}{2\pi}\oint \mathbf{p}\cdot \mathbf{ds}. \tag{1.15}$$

The corresponding adiabatic invariant is

$$J = 2pI, \tag{1.16}$$

$$I = \int_{s'_m}^{s_m} \cos\alpha ds = \int_{s'_m}^{s_m}\left(1 - \frac{B(s)}{B_m}\right)^{\frac{1}{2}} ds. \tag{1.17}$$

In the dipole field, the numerical integration for I is approximated as (Schulz and Lanzerotti 1974)

$$I = LR_E Y(\alpha_e), \tag{1.18}$$

$$Y(\alpha_e) = 2.760346 + 2.357194\sin\alpha_e - 5.117540\sin^{3/4}\alpha_e. \tag{1.19}$$

The third periodic motion is the drift circling the Earth. Roederer (1970) had given the bounce-averaged drift velocity in the absence of electric field

$$\langle \mathbf{v_d}\rangle = \frac{2p}{q\tau_b B_e^2}\nabla_e \mathbf{I}\times \mathbf{B_e}, \tag{1.20}$$

with the subscript e denoting the values at the equator. In the dipole field, this drift velocity only has the component along the longitudinal direction, and the bounce-averaged drift frequency is written as (Roederer 1970; Schulz and Lanzerotti 1974)

$$\omega_d = \frac{3m_0c^2\gamma\beta^2 L}{|q|B_0R_E^2}\frac{D(\alpha_e)}{T(\alpha_e)}, \tag{1.21}$$

with the equatorial magnetic field $B_0 = 31200$ nT at the surface of Earth, and an approximation function

$$D(\alpha_e) = \frac{1}{12}\left(5.520692 - 2.357194\sin\alpha_e + 1.279385\sin^{3/4}\alpha_e\right). \tag{1.22}$$

The drift period is

$$\tau_d = \frac{2\pi}{\omega_d} = \frac{2\pi|q|B_0R_E^2}{3m_0c^2\gamma\beta^2 L}\frac{T(\alpha_e)}{D(\alpha_e)}. \tag{1.23}$$

The corresponding action variable is

$$J_3 = \frac{1}{2\pi}\oint(\mathbf{p} + q\mathbf{A})\cdot d\mathbf{l} \approx \frac{q}{2\pi}\Phi, \tag{1.24}$$

with the total magnetic flux Φ enclosed by the drift trajectory. In the dipole field, the magnetic flux is

$$\Phi = \frac{2\pi B_0R_E^2}{L}. \tag{1.25}$$

In the arbitrary magnetic field, Roederer (1970) had defined the generalized magnetic shell L^*

$$L^* = \frac{2\pi B_0R_E^2}{\Phi}, \tag{1.26}$$

which is the third adiabatic invariant.

1.3.2 Kinetic Theory

The distribution function $f(t, \mathbf{r}, \mathbf{p})$ in the six-dimensional phase space (r, p) satisfies the Vlasov equation

$$\frac{\partial f}{\partial t} + \frac{d\mathbf{r}}{dt}\cdot\frac{\partial f}{\partial \mathbf{r}} + \frac{d\mathbf{p}}{dt}\cdot\frac{\partial f}{\partial \mathbf{p}} = 0. \tag{1.27}$$

$f(t, \mathbf{r}, \mathbf{p})$ is defined as the number density in the unit phase space (Lyons and Williams 1984)

$$f = \frac{\Delta N}{\Delta x \Delta y \Delta z \Delta p_x \Delta p_y \Delta p_z}. \tag{1.28}$$

The differential flux j is defined as the number of particles at the energy E_k within ΔE_k, coming from the direction Ω within a solid angle $\Delta\Omega$, crossing a unit area ΔA perpendicular to Ω within a time interval Δt (Roederer 1970)

$$j = \frac{\Delta N}{\Delta t \Delta A \Delta \Omega \Delta E_k}. \tag{1.29}$$

Their relation can be written as (Lyons and Williams 1984)

$$j = p^2 f. \tag{1.30}$$

In the action-angle space, the Vlasov equation for phase space density (PSD) $f(t, \mathbf{J}, \boldsymbol{\varphi})$ evolution can be expressed as (Boscher et al. 1996)

$$\frac{\partial f}{\partial t} + \frac{\mathrm{d}\mathbf{J}}{\mathrm{d}t} \cdot \frac{\partial f}{\partial \mathbf{J}} + \frac{\mathrm{d}\boldsymbol{\varphi}}{\mathrm{d}t} \cdot \frac{\partial f}{\partial \boldsymbol{\varphi}} = 0. \tag{1.31}$$

If all the three adiabatic invariants are conserved $\mathrm{d}\mathbf{J}/\mathrm{d}t = 0$, the φ-averaged PSD $f(t, \mathbf{J})$ satisfies

$$\frac{\partial f}{\partial t} = 0, \tag{1.32}$$

representing the conservation of PSD. There exist many stochastic processes violating the adiabatic invariants, which are often characterized as the diffusion-type processes (Beutier and Boscher 1995; Boscher et al. 1996; Bourdarie et al. 1997)

$$\frac{\partial f}{\partial t} = \sum_{i=1}^{3} \sum_{j=1}^{3} \frac{\partial}{\partial J_i} \left(D_{ij} \frac{\partial f}{\partial J_j} \right). \tag{1.33}$$

The diffusion coefficients $D_{i,j}$ are defined as (Glauert and Horne 2005)

$$D_{ij} = \frac{1}{2} \left\langle \frac{\Delta J_i \Delta J_j}{\Delta t} \right\rangle \tag{1.34}$$

with $\left\langle \frac{\Delta J_i \Delta J_j}{\Delta t} \right\rangle$ denoting the expected value per unit time of the changes $\Delta J_i \Delta J_j$.

In the arbitrary phase space $\boldsymbol{\zeta} = (\zeta_1, \zeta_2, \zeta_3)$, the PSD $f(t, \boldsymbol{\zeta})$ evolution is described by (Schulz and Lanzerotti 1974; Walt 1994)

$$\frac{\partial f(t, \boldsymbol{\zeta})}{\partial t} = \sum_{i=1}^{3}\sum_{j=1}^{3} \frac{1}{\Im} \frac{\partial}{\partial \zeta_i} \left[D_{ij} \Im \frac{\partial f(t, \zeta)}{\partial \zeta_j} \right], \tag{1.35}$$

with the Jacobian matrix $\Im$

$$\Im = \frac{\partial(J_1, J_2, J_3)}{\partial(\zeta_1, \zeta_2, \zeta_3)}, \tag{1.36}$$

and diffusion coefficients D_{ij}

$$D_{ij} = \frac{1}{2} \left\langle \frac{\Delta\zeta_i \Delta\zeta_j}{\Delta t} \right\rangle. \tag{1.37}$$

The PSD $f(t, \mathbf{J}, \varphi_3)$ averaged over the first two angle variables obey the following convection-diffusion equation (Boscher et al. 1996; Bourdarie et al. 1997)

$$\frac{\partial f}{\partial t} + \frac{\mathrm{d}J_3}{\mathrm{d}t} \frac{\partial f}{\partial J_3} + \frac{\mathrm{d}\varphi_3}{\mathrm{d}t} \frac{\partial f}{\partial \varphi_3} = \sum_{i=1}^{3}\sum_{j=1}^{3} \frac{\partial}{\partial J_i} \left(D_{ij} \frac{\partial f}{\partial J_j} \right). \tag{1.38}$$

Compared to the previous diffusion equation (1.33), the current convection-diffusion equation has higher spatial and temporal resolutions (Boscher et al. 1996).

1.4 Research Contents

The radiation belt discovered in 1958 has received increasing attention in the space physics community (Friedel et al. 2002; Thorne 2010). The radiation belt research has both scientific and practical significance. From the scientific point of view, there exist some basic physical questions that remain to be answered, for example, why the electron radiation belt responds so differently to different magnetic storms; what mechanisms control the buildup or dropout of radiation belt electrons. From the practical point of view, the energetic radiation belt particles can significantly harm the space-borne systems, and the interruption of service (e.g., navigation and weather forecasting) provided by the space-borne systems will seriously affect the production and living of modern humans. Radiation belt models are required to understand the physical mechanisms for its sources and sinks, and predict its variability. The test-particle and kinetic formulations are two widely used approaches. This thesis concentrates on the development and application of radiation belt kinetic models.

References

Albert JM, Meredith NP, Horne RB (2009) Three-dimensionaldiffusion simulation of outer radiation belt electrons during the october 9 1990, magnetic storm. J Geophys Res 114(A09):214. doi: 10.1029/2009JA014336

Alfven H (1950) Cosmical electrodynamics. Oxford University Press, London

Baker DN (2002) How to cope with space weather. Science 297(5586):1486

Baker DN, Klebesadel RW, Higbie PR, Blake JB (1986) Highly relativistic electrons in the Earth's outer magnetosphere. I -Lifetimes and temporal history 1979–1984. J Geophys Res 91: 4265–4276. doi:10.1029/JA091iA04p04265

Baker DN, Blake JB, Callis LB, Belian RD, Cayton TE (1989) Relativistic electrons near geostationary orbit—Evidence for internal magnetospheric acceleration. Geophys Res Lett 16:559–562. doi:10.1029/GL016i006p00559

Baker DN, Kanekal S, Blake JB, Klecker B, Rostoker G (1994) Satellite anomalies linked to electron increase in the magnetosphere. EOS Trans 75:401–405. doi:10.1029/94EO01038

Beutier T, Boscher D (1995) A three-dimensional analysis of the electron radiation belt by the salammbô code. J Geophys Res 100:14853–14862. doi:10.1029/94JA03066

Bortnik J, Thorne RM, O'Brien TP, Green JC, Strangeway RJ, Shprits YY, Baker DN (2006) Observation of two distinct, rapid loss mechanisms during the 20 November 2003 radiation belt dropout event. J Geophys Res 111(A12):216. doi:10.1029/2006JA011802

Boscher D, Bourdarie S, Beutier T (1996) Dynamic modeling of trapped particles. IEEE Trans Nucl Sci 43:416–425. doi:10.1109/23.490890

Bourdarie S, Boscher D, Beutier T, Sauvaud J, Blanc M (1997) Electron and proton radiation belt dynamic simulations during storm periods: a new asymmetric convection-diffusion model. J Geophys Res 102:17541–17552. doi:10.1029/97JA01305

Brautigam DH, Albert JM (2000) Radial diffusion analysis of outer radiation belt electrons during the October 9, 1990, magnetic storm. J Geophys Res 105:291–310. doi:10.1029/1999JA900344

Friedel RHW, Reeves GD, Obara T (2002) Relativistic electron dynamics in the inner magnetosphere—a review. J Atmos Sol-Terr Phys 64:265–282. doi:10.1016/S1364-6826(01)00088-8

Glauert SA, Horne RB (2005) Calculation of pitch angle andenergy diffusion coefficients with the PADIE code. J Geophys Res 110(A04):206. doi:10.1029/2004JA010851

Hamlin DA, Karplus R, Vik RC, Watson KM (1961) Mirrorand azimuthal drift frequencies for geomagnetically trapped particles. J Geophys Res 66:1–4. doi:10.1029/JZ066i001p00001

Kim H-J, Chan AA (1997) Fully adiabatic changes in storm time relativistic electron fluxes. J Geophys Res 102:22107–22116. doi:10.1029/97JA01814

Konradi A, Semar CL, Fritz TA (1975) Substorm-injected protons and electrons and the injection boundary model. J Geophys Res 80:543–552. doi:10.1029/JA080i004p00543

Li X, Baker DN, Kanekal SG, Looper M, Temerin M (2001) Long term measurements of radiation belts by SAMPEX and their variations. Geophys Res Lett 28:3827–3830. doi:10.1029/2001GL013586

Lyons LR, Thorne RM (1973) Equilibrium structure of radiationbelt electrons. J Geophys Res 78:2142–2149. doi:10.1029/JA078i013p02142

Lyons LR, Williams DJ (1984) Quantitative aspects of magnetospheric physics. Springer, New York

Lyons LR, Thorne RM, Kennel CF (1972) Pitch-angle diffusion of radiation belt electrons within the plasmasphere. J Geophys Res 77:3455–3474. doi:10.1029/JA077i019p03455

Mauk BH, McIlwain CE (1974) Correlation of K_p with the substorm-injected plasma boundary. J Geophys Res 79:3193–3196. doi:10.1029/JA079i022p03193

Mauk BH, Meng C-I (1983) Characterization of geostationary particle signatures based on the 'injection boundary' model. J Geophys Res 88:3055–3071. doi:10.1029/JA088iA04p03055

McIlwain CE (1974) Substorm injection boundaries. In: McCormac BM (ed) Magnetospheric physics. pp. 143–154

Northrop T (1963) The Adiabatic motion of charged particles. Wiley (Interscience), New York

Paulikas GA, Blake JB (1979) Effects of the solar wind on magnetospheric dynamics—energetic electrons at the synchronous orbit. American Geophysical Union Geophysical Monograph Series, Washington DC, pp 180–202

Reeves GD, Henderson MG, McLachlan PS, Belian RD, Friedel RHW, Korth A (1996) Radial propagation of substorm injections. In: Rolfe EJ, Kaldeich B (eds) International conference on substorms, vol 389. ESA Special Publication, pp. 579–584

Reeves GD, McAdams KL, Friedel RHW, O'Brien TP (2003) Acceleration and loss of relativistic electrons during geomagnetic storms. Geophys Res Lett 30:1529. doi:10.1029/2002GL016513

Roederer JG (1970) Dynamics of geomagnetically trapped radiation. Springer, New York

Schulz M, Lanzerotti LJ (1974) Particle diffusion in the radiation belts, in physics and chemistry in space, vol 7. Springer, New York

Spitzer L (1965) Physics of fully ionized gases. Wiley (Interscience), New York

Su Z, Xiao F, Zheng H, Wang S (2011) CRRES observation and STEERB simulation of the 9 October 1990 electron radiation belt dropout event. Geophys Res Lett 38(L06):106. doi: 10.1029/2011GL046873

Thorne RM (2010) Radiation belt dynamics: the importance of wave-particle interactions. Geophys Res Lett 37(L22):107. doi:10.1029/2010GL044990

Van Allen JA, Frank LA (1959) Radiation around the Earth to a radial distance of 107,400 km. Nature 183:430–434. doi:10.1038/183430a0

Walt M (1994) Introduction to geomagnetically trapped radiation. Cambridge University Press, Cambridge

Chapter 2
Local Diffusion

2.1 Introduction

Trapped particles exhibit three periodic motions (see Sect. 1.3.1), and consequently possess three adiabatic invariants. Radiation belt electrons can cyclotron resonate with some magnetospheric plasma waves (violating the first and second adiabatic invariants), and then experience the significant acceleration or loss process (Horne and Thorne 1998, 2003; Summers et al. 1998; Roth et al. 1999; Summers and Ma 2000; Albert 2002, 2003, 2004, 2005; Horne et al. 2003a, b, 2005; O'Brien et al. 2003; Albert and Young 2005; Li et al. 2005; Varotsou et al. 2005, 2008; Shprits et al. 2006, 2009b; Xiao and He 2006; Xiao et al. 2007a, b, 2009, 2010a, b; Li et al. 2007; Su and Zheng 2008, 2009; Zheng et al. 2008; Albert and Bortnik 2009; Albert et al. 2009; Su et al. 2009a, c, 2010a, c, 2011a, b, c; Thorne 2010). Such cyclotron resonance driven by various waves is widely believed to control the radiation belt electron dynamics.

The whistler-mode chorus waves are usually distributed in the low-density plasmatrough region approximately from midnight through dawn to noon with frequencies $0.05-0.8\,|\Omega_e|$ ($|\Omega_e|$ is the equatorial electron gyro-frequency) (Tsurutani and Smith 1974, 1977; Koons and Roeder 1990; Meredith et al. 2001; Santolík et al. 2003, 2004). Typical chorus amplitudes are found to be 1–100 pT (Burtis and Helliwell 1975; Meredith et al. 2001, 2003), and even approach 1–3 nT during geomagnetically active periods (Parrot and Gaye 1994; Cattell et al. 2008; Cully et al. 2008). The chorus waves are widely considered to cause the significant acceleration of radiation belt electrons on a timescale of days (e.g., Horne and Thorne 1998; Summers et al. 1998; Horne and Thorne 2003).

The whistler-mode hiss waves mainly occur in the high-density plasmasphere and plasmaspheric plume (Russell et al. 1969; Thorne et al. 1973; Li et al. 2007; Bortnik et al. 2008b, 2009). Their frequencies range from ~100 Hz to several kHz, with amplitudes 10 pT during the quiet times and ~100 pT during the storm times (Smith et al. 1974; Meredith et al. 2004; Summers et al. 2008). The hiss waves can

Z. Su, *A Global Kinetic Model for Electron Radiation Belt Formation and Evolution*, Springer Theses, DOI 10.1007/978-3-662-46651-3_2

scatter the energetic electrons into the loss cone and result in their slow decay on a timescale of several days (Lyons and Thorne 1973; Meredith et al. 2007; Su et al. 2010a, 2011a, b, c).

The electromagnetic ion cyclotron (EMIC) waves with typical frequencies 0.1–5.0 Hz are observed in the plasmasphere along the duskside plasmapause or plasmaspheric plumes (Fraser et al. 1996). EMIC emissions typically split into three distinct bands below the hydrogen (H^+), helium (He^+), and oxygen (O^+) ion gyrofrequencies in a multi-ion (H^+, He^+, O^+) plasma. The EMIC waves can efficiently cause the precipitation loss of relativistic electrons on a timescale of hours (Summers et al. 2007a, b).

The quasilinear diffusion theory is widely used to describe the electron PSD evolution due to resonance with waves (e.g., Kennel and Engelmann 1966; Schulz and Lanzerotti 1974; Beutier et al. 1995; Bourdarie et al. 1997; Albert and Young 2005; Xiao et al. 2009; Su et al. 2010a, 2011b). The detailed expressions for the relativistic quasilinear diffusion have been given by Albert (2003, 2005), Glauert and Horne (2005), and their corresponding parallel propagation approximation (PPA) and mean value approximation (MVA) have been proposed by Summers (2005) and Albert (2008). In this chapter, we develop a local diffusion model for radiation belt electron evolution due to cyclotron resonance with various plasma waves. Using this numerical model, we examine the effect of chorus, hiss, and EMIC waves on the radiation belt electrons. Furthermore, in view of ignoring of cross-pitch-angle-momentum diffusion in the previous models, we here evaluate the importance of cross term in the simulations.

2.2 Local Radiation Belt Diffusion Model

2.2.1 Background Magnetic Field

Similar to previous works (e.g., Albert 2003, 2005, 2008; Horne et al. 2003a, 2005; Glauert and Horne 2005; Summers 2005), a typical dipole magnetic field model is used to describe the inner magnetospheric field configuration. It should be mentioned that, during geomagnetic storms, the diffusion coefficients in the different geomagnetic field models may have quantitative differences (Orlova and Shprits 2010).

2.2.2 Basic Equation

The Jacobian matrix between the action space (J_1, J_2, J_3) and observable space (α_e, p, L) can be written as

$$
\begin{aligned}
\Im &\propto \begin{vmatrix} \partial\mu/\partial\alpha_e & \partial\mu/\partial p & \partial\mu/\partial L \\ \partial J/\partial\alpha_e & \partial J/\partial p & \partial J/\partial L \\ 0 & 0 & \mathrm{d}\Phi/\mathrm{d}L \end{vmatrix} \\
&\propto \frac{\mathrm{d}\Phi}{\mathrm{d}L}\left(\frac{\partial\mu}{\partial\alpha_e}\frac{\partial J}{\partial p} - \frac{\partial\mu}{\partial p}\frac{\partial J}{\partial\alpha_e}\right) \\
&\propto Lp^2 \sin\alpha_e \cos\alpha_e \left(I - \frac{\sin\alpha_e}{\cos\alpha_e}\frac{\partial I}{\partial\alpha_e}\right).
\end{aligned}
\tag{2.1}
$$

From Eq. (1.17), we can obtain

$$
\frac{\partial I}{\partial\alpha_e} = -\int_{s'_m}^{s_m} \sin\alpha \frac{\partial\alpha}{\partial\alpha_e}\mathrm{d}s + \cos\alpha(s_m)\frac{\partial s_m}{\partial\alpha_e} - \cos\alpha(s'_m)\frac{\partial s'_m}{\partial\alpha_e}. \tag{2.2}
$$

Because of $\alpha(s_m) = \alpha(s'_m) = 90°$, the equation above is simplified as

$$
\frac{\partial I}{\partial\alpha_e} = -\int_{s'_m}^{s_m} \sin\alpha \frac{\partial\alpha}{\partial\alpha_e}\mathrm{d}s. \tag{2.3}
$$

Considering the conservation of first adiabatic invariant in the course of bounce motion

$$
\frac{\sin^2\alpha}{B} = \frac{\sin^2\alpha_e}{B_e}, \tag{2.4}
$$

we can derive

$$
\frac{\partial\alpha}{\partial\alpha_e} = \frac{\tan\alpha}{\tan\alpha_e}. \tag{2.5}
$$

Substitution of (2.5) into (2.3) yields

$$
I - \frac{\sin\alpha_e}{\cos\alpha_e}\frac{\partial I}{\partial\alpha_e} = \int_{s'_m}^{s_m} \cos\alpha \mathrm{d}s + \int_{s'_m}^{s_m} \frac{\sin^2\alpha}{\cos\alpha}\mathrm{d}s = \int_{s'_m}^{s_m} \frac{1}{\cos\alpha}\mathrm{d}s. \tag{2.6}
$$

We further substitute (2.6), (1.12) and (1.14) into (2.1), and obtain the simplified Jacobian matrix

$$
\Im \propto L^2 G, \tag{2.7}
$$

$$
G = p^2 T(\alpha_e) \sin\alpha_e \cos\alpha_e. \tag{2.8}
$$

The cyclotron resonance between radiation belt electrons and various plasma waves is a "local" physical process violating the first and second adiabatic invariants (Schulz and Lanzerotti 1974; Shprits et al. 2008a, b). Substitution of the Jacobian matrix (2.7) into (1.35) yields the quasilinear diffusion equation for electron PSD evolution (Schulz and Lanzerotti 1974; Lyons and Williams 1984; Kozyra et al. 1994)

$$\begin{aligned}\frac{\partial f}{\partial t} &= \frac{1}{Gp}\frac{\partial}{\partial \alpha_e}\left[G\left(\langle D_{\alpha\alpha}\rangle \frac{1}{p}\frac{\partial f}{\partial \alpha_e} + \langle D_{\alpha p}\rangle \frac{\partial f}{\partial p}\right)\right] \\ &+ \frac{1}{G}\frac{\partial}{\partial p}\left[G\left(\langle D_{p\alpha}\rangle \frac{1}{p}\frac{\partial f}{\partial \alpha_e} + \langle D_{pp}\rangle \frac{\partial f}{\partial p}\right)\right]\end{aligned} \tag{2.9}$$

with the bounce-averaged pitch-angle, momentum, and cross diffusion coefficients (Lyons and Williams 1984; Glauert and Horne 2005)

$$\langle D_{\alpha\alpha}\rangle = \frac{p^2}{2}\left\langle \frac{(\Delta\alpha_e)^2}{\Delta t}\right\rangle, \tag{2.10}$$

$$\langle D_{\alpha p}\rangle = \frac{p}{2}\left\langle \frac{\Delta\alpha_e \Delta p}{\Delta t}\right\rangle, \tag{2.11}$$

$$\langle D_{pp}\rangle = \frac{1}{2}\left\langle \frac{(\Delta p)^2}{\Delta t}\right\rangle. \tag{2.12}$$

2.2.3 *Diffusion Coefficients*

The bounce-averaged diffusion coefficients $\langle D_{\alpha\alpha}\rangle$, $\langle D_{pp}\rangle$ and $\langle D_{\alpha p}\rangle = \langle D_{p\alpha}\rangle$ can be obtained through integration of local diffusion coefficients $D_{\alpha\alpha}$, D_{pp} and $D_{\alpha p} = D_{p\alpha}$ along the geomagnetic field line. In the dipole magnetic field, their expressions can be written as (Lyons and Williams 1984)

$$\langle D_{\alpha\alpha}\rangle = \frac{1}{T}\int_0^{\lambda_m} D_{\alpha\alpha}\frac{\cos\alpha}{\cos^2\alpha_e}\cos^7\lambda d\lambda, \tag{2.13}$$

$$\langle D_{pp}\rangle = \frac{1}{T}\int_0^{\lambda_m} D_{pp}\frac{(1+3\sin^2\lambda)^{1/2}}{\cos\alpha}\cos\lambda d\lambda, \tag{2.14}$$

$$\langle D_{\alpha p}\rangle = \langle D_{p\alpha}\rangle = \frac{1}{T}\int_0^{\lambda_m} D_{\alpha p}\frac{(1+3\sin^2\lambda)^{1/4}}{\cos\alpha_e}\cos^4\lambda d\lambda, \tag{2.15}$$

with the latitude λ and the mirror latitude λ_m corresponding to the equatorial pitch-angle α_e. Based on the conservation of first adiabatic invariant and the expression of dipole magnetic field, we can obtain the relation between α_e and λ_m

$$\frac{1}{\sin^2 \alpha_e} = \frac{\sin^2 \alpha(\lambda_m)}{\sin^2 \alpha_e} = \frac{B(\lambda_m)}{B_e} = \frac{(1 + 3\sin^2 \lambda_m)^{1/2}}{\cos^6 \lambda_m}, \tag{2.16}$$

$$\cos^{12} \lambda_m + 3\cos^2 \lambda_m \sin^4 \alpha_e - 4\sin^4 \alpha_e = 0. \tag{2.17}$$

The calculation of local diffusion coefficients is implemented through three steps:

1. Determination of wave spectral distribution. The wave frequency ω is usually assumed to obey the typical Gaussian distribution (e.g., Albert 2003, 2005; Summers 2005; Glauert and Horne 2005)

$$B_\omega^2 = \begin{cases} A^2 \exp[-(\omega - \omega_m)^2/(\delta\omega)^2], & \omega_1 \le \omega \le \omega_2 \\ 0, & \text{other} \end{cases}, \tag{2.18}$$

$$A^2 = \frac{2B_t^2}{\pi^{1/2}\delta\omega}\left[\operatorname{erf}\left(\frac{\omega_2 - \omega_m}{\delta\omega}\right) + \operatorname{erf}\left(\frac{\omega_m - \omega_1}{\delta\omega}\right)\right]^{-1}, \tag{2.19}$$

with the lower limit ω_1, upper limit ω_2, center ω_m, half-width $\delta\omega$, and amplitude B_t. The tangent X of wave normal angle θ is also assumed to obey the typical Gaussian distribution (e.g., Albert 2003, 2005; Summers 2005; Glauert and Horne 2005)

$$g(X) = \begin{cases} \exp[-(X - X_m)^2/X_\omega^2], & X_1 \le X \le X_2 \\ 0, & \text{other} \end{cases}, \tag{2.20}$$

with the lower limit X_1, upper limit X_2, center X_m and half-width X_ω. When $X_m = X_\omega = 0$, the waves propagate fully parallel along the magnetic field lines.

2. Calculation of resonance roots. The resonance frequency ω and wave vector k satisfy the dispersion relation (Stix 1992) and the n-order cyclotron resonance condition

$$\omega - v_\parallel k_\parallel = -n|\Omega_e|/\gamma, \tag{2.21}$$

where $v_\parallel = v\cos\alpha$ and $k_\parallel = k\cos\theta$ represent the parallel component of electron velocity and wave vector with respect to the background magnetic field. Under high-density approximation (i.e., the ratio of electron plasma frequency ω_{pe} to electron gyrofrequency $|\Omega_e|$ is much larger than 1), the dispersion relation of whistler-mode chorus wave in the electron-proton plasma can be simplified as (Lyons 1974c)

$$V^2 = \left(\frac{ck}{\omega}\right)^2 = \frac{\omega_{pe}^2}{|\Omega_e|^2}\frac{1+M}{M}\Psi^{-1}, \tag{2.22}$$

$$\Psi = 1 - \frac{\omega^2}{\Omega_p|\Omega_e|} - \frac{\sin^2\theta}{2} + \left[\frac{\sin^4\theta}{4} + \frac{\omega^2}{\Omega_p^2}(1-M)^2\cos^2\theta\right]^{1/2}, \tag{2.23}$$

with the ratio of electron and proton masses $M = m_e/m_p = 1/1836$ and the proton gyrofrequency Ω_p. In a multi-component plasma (e^-, H^+, He^+, O^+), the dispersion relation of parallel-propagation EMIC waves is (Summers and Thorne 2003)

$$\frac{c^2k^2}{\omega^2} = 1 - \frac{\omega_{pe}^2}{\omega(\omega + |\Omega_e|)} - \sum_{j=1}^{3} \frac{\omega_{pj}^2}{\omega(\omega - \Omega_j)}, \tag{2.24}$$

where $\Omega_j|_{j=1,2,3}$ represent the gyro-frequencies of H^+, He^+ and O^+, and ω_{pj} denote the jth ion plasma frequency.

3. Expression of diffusion coefficients. The quasilinear description of cyclotron resonance was first proposed by Kennel and Engelmann (1966). Based on the works of Kennel and Engelmann (1966), Lyons (1974a, b, c), Lyons and Williams (1984), Albert (2003, 2005), Glauert and Horne (2005) have given the expression of relativistic diffusion coefficients for the obliquely propagating plasma waves

$$\begin{pmatrix} D_{\alpha\alpha} & D_{\alpha p} \\ D_{p\alpha} & D_{pp} \end{pmatrix} = \begin{pmatrix} \tilde{D}_{\alpha\alpha} & \tilde{D}_{\alpha p} \\ \tilde{D}_{p\alpha} & \tilde{D}_{pp} \end{pmatrix} |\Omega_e|^2 \frac{B_\omega^2}{B^2} \frac{p^2}{\gamma^2}, \tag{2.25}$$

$$\begin{pmatrix} \tilde{D}_{\alpha\alpha} & \tilde{D}_{\alpha p} \\ \tilde{D}_{p\alpha} & \tilde{D}_{pp} \end{pmatrix} = \sum_{n=n_1}^{n_2} \int_{X_1}^{X_2} X\mathrm{d}X \begin{pmatrix} \tilde{D}_{\alpha\alpha}^{nX} & \tilde{D}_{\alpha p}^{nX} \\ \tilde{D}_{p\alpha}^{nX} & \tilde{D}_{pp}^{nX} \end{pmatrix}, \tag{2.26}$$

$$\tilde{D}_{\alpha\alpha}^{nX} = \sum_i \frac{\omega^2(\sin^2\alpha + n|\Omega_e|/\gamma\omega)^2}{4\pi v_\parallel^2(1+X^2)N(\omega)} \left. \frac{g(X)|\Phi_{n,\mathbf{k}}|^2}{\left|v_\parallel - \partial\omega/\partial k_\parallel|_X\right|} \right|_{\substack{\omega=\omega_i \\ k=k_i}}, \tag{2.27}$$

$$N(\omega) = \frac{1}{2\pi^2} \int_0^\infty g(X) \left| J \begin{pmatrix} k_\perp, k_\parallel \\ \omega, X \end{pmatrix} \right| k_\perp \mathrm{d}X, \tag{2.28}$$

$$J \begin{pmatrix} k_\perp, k_\parallel \\ \omega, X \end{pmatrix} = -k_\parallel \left. \frac{\partial k_\parallel}{\partial\omega} \right|_X = -k\cos^2\theta \left. \frac{\partial k}{\partial\omega} \right|_X, \tag{2.29}$$

$$\begin{aligned} |\Phi_{n,\mathbf{k}}|^2 &= \left[\left(\frac{2D}{V^2 - S} \right)^2 + \left(\frac{2P\cos\theta}{V^2\sin^2\theta - P} \right)^2 \right]^{-1} \\ &\times \left[\left(\frac{V^2 - L}{V^2 - S} \right) J_{n+1} + \left(\frac{V^2 - R}{V^2 - S} \right) J_{n-1} + \frac{V^2\cot\alpha\sin 2\theta}{V^2\sin^2\theta - P} J_n \right]^2, \end{aligned} \tag{2.30}$$

$$R = \frac{\omega_{pe}^2}{|\Omega_e|^2} \frac{|\Omega_e|}{\omega} \left[\frac{1+M}{1 - M - (\omega/|\Omega_e| - \Omega_p/\omega)} \right], \tag{2.31}$$

$$L = -\frac{\omega_{pe}^2}{|\Omega_e|^2}\frac{|\Omega_e|}{\omega}\left[\frac{1+M}{1-M+(\omega/|\Omega_e|-\Omega_p/\omega)}\right], \tag{2.32}$$

$$P = -\frac{\omega_{pe}^2}{|\Omega_e|^2}\frac{|\Omega_e|^2}{\omega^2}(1+M), \tag{2.33}$$

$$S = \frac{R+L}{2}, \tag{2.34}$$

$$D = \frac{R-L}{2}, \tag{2.35}$$

$$\tilde{D}_{\alpha p}^{nX} = \tilde{D}_{p\alpha}^{nX} = \tilde{D}_{\alpha\alpha}^{nX}\left[\frac{\sin\alpha\cos\alpha}{-\sin^2\alpha - n|\Omega_e|/\gamma\omega}\right]\Bigg|_{\substack{\omega=\omega_i\\ k=k_i}}, \tag{2.36}$$

$$\tilde{D}_{pp}^{nX} = \tilde{D}_{\alpha\alpha}^{nX}\left[\frac{\sin\alpha\cos\alpha}{-\sin^2\alpha - n|\Omega_e|/\gamma\omega}\right]^2\Bigg|_{\substack{\omega=\omega_i\\ k=k_i}}. \tag{2.37}$$

Here n_1 and n_2 are the upper and lower limits of resonance orders; ω_i and k_i are the i-th resonant frequency and wave number, the subscripts $\perp$ and $\|$ represent the parallel and perpendicular components with respect to the background magnetic field; the argument of n-order Bessel function is $-k_\perp p_\perp/(m_e|\Omega_e|)$. Based on the work of Kennel and Engelmann (1966), Summers (2005) and Tao et al. (2011a) had presented the expressions of relativistic diffusion coefficients for the parallel propagation plasma waves

$$D_{\alpha\alpha} = |\Omega_e|^2\left(\frac{p^2}{\gamma^2}I_0 - 2\cos\alpha\frac{m_e c p}{\gamma}I_1 + \cos^2\alpha m_e^2 c^2 I_2\right), \tag{2.38}$$

$$D_{pp} = m_e^2 c^2 |\Omega_e|^2 \sin^2\alpha I_2, \tag{2.39}$$

$$D_{\alpha p} = D_{p\alpha} = -m_e c p|\Omega_e|^2 \sin\alpha\left(\frac{I_1}{\gamma} - \frac{m_e c\cos\alpha}{p}I_2\right), \tag{2.40}$$

$$I_s = \frac{\pi}{4}\sum_i\left[\frac{B_\omega^2}{B^2}\left(\frac{\omega}{ck}\right)^s\left|1-\cos\alpha\frac{p}{\gamma m_e}\frac{\mathrm{d}k}{\mathrm{d}\omega}\right|^{-1}\right]\Bigg|_{\substack{\omega=\omega_i\\ k=k_i}}, \quad s = 0, 1, 2. \tag{2.41}$$

2.2.4 Numerical Method

The quasilinear diffusion equation includes three parts: pitch-angle diffusion, momentum diffusion, and cross diffusion. Utilizing the standard finite difference methods to this equation often leads to the numerical instability associated with the cross diffusion (Albert 2004; Albert and Young 2005). Albert and Young (2005) proposed a variable transformation method to diagonalize the diffusion matrix. In this special variable space, cross diffusion is removed and the typical implicit method can be directly used to solve the equation. Obviously, such transformation depends on the diffusion coefficients, and the different diffusion coefficients would require different transformation variables. Tao et al. (2008, 2009) adopted a Monte Carlo method or a layer method to solve this equation. But these methods are inefficient compared to the direct difference method. We have developed a Hybrid Finite Difference Method (Xiao et al. 2009; Su et al. 2009c, 2011c) to fully solve this quasilinear diffusion Eq. (2.9).

The diffusion Eq. (2.9) is rewritten as

$$\begin{aligned}\frac{\partial f}{\partial t} &= \frac{1}{Gp}\frac{\partial}{\partial \alpha_e}\left(\frac{G\langle D_{\alpha\alpha}\rangle}{p}\frac{\partial f}{\partial \alpha_e}\right) + \frac{1}{Gp}\frac{\partial}{\partial \xi}\left(\frac{G\langle D_{pp}\rangle}{p}\frac{\partial f}{\partial \xi}\right)\\ &+ \frac{1}{Gp}\frac{\partial}{\partial \alpha_e}\left(\frac{G\langle D_{\alpha p}\rangle}{p}\right)\frac{\partial f}{\partial \xi} + \frac{1}{Gp}\frac{\partial}{\partial \xi}\left(\frac{G\langle D_{\alpha p}\rangle}{p}\right)\frac{\partial f}{\partial \alpha_e} + \frac{2\langle D_{\alpha p}\rangle}{p^2}\frac{\partial^2 f}{\partial \alpha_e \partial \xi},\end{aligned} \tag{2.42}$$

with the variable ξ defined as $\xi = \ln \frac{p}{m_e c}$. Based on the split-operator technique (Strang 1968; Kim et al. 1999), the equation above is split into

$$\frac{\partial f}{\partial t} = \frac{1}{Gp}\frac{\partial}{\partial \alpha_e}\left(\frac{G\langle D_{\alpha\alpha}\rangle}{p}\frac{\partial f}{\partial \alpha_e}\right), \tag{2.43}$$

$$\frac{\partial f}{\partial t} = \frac{1}{Gp}\frac{\partial}{\partial \xi}\left(\frac{G\langle D_{pp}\rangle}{p}\frac{\partial f}{\partial \xi}\right), \tag{2.44}$$

$$\frac{\partial f}{\partial t} = \frac{1}{Gp}\frac{\partial}{\partial \xi}\left(\frac{G\langle D_{\alpha p}\rangle}{p}\right)\frac{\partial f}{\partial \alpha_e} + \frac{1}{Gp}\frac{\partial}{\partial \alpha_e}\left(\frac{G\langle D_{\alpha p}\rangle}{p}\right)\frac{\partial f}{\partial \xi} + \frac{2\langle D_{\alpha p}\rangle}{p^2}\frac{\partial^2 f}{\partial \alpha_e \partial \xi}. \tag{2.45}$$

At each time-step, the solving of Eq. (2.42) can be implemented by the successive solving of Eqs. (2.43–2.45). The diagonal Eqs. (2.43) and (2.44) are solved by the fully-implicit method, while the off-diagonal Eq. (2.45) is solved by the alternative direction implicit (ADI) method (Strang 1968). The whole numerical algorithm is named as the Hybrid Finite Difference Method (HFD) due to the combination of two different finite difference methods. In the quasilinear dynamic simulations of radiation belt, ring current and aurora, this HFD method has been widely used by our group (Su et al. 2009a, b, c, 2010a, b, c, 2011a, b, c; Xiao et al. 2009, 2010a, b, 2011) and other researchers (Fok et al. 2010; Thorne et al. 2010; Tao et al. 2011b; Zheng et al. 2011).

2.3 Idealized Simulations

Because of the numerical difficulty associated with cross diffusion, the previous work often ignore the cross diffusion terms (e.g., Varotsou et al. 2005, 2008; Li et al. 2007). In this section, we calculate the diffusion coefficients of chorus, hiss, and EMIC waves, and then fully solve the diffusion equation to quantify the effect of various waves on radiation belt electrons. In addition, through comparison between simulations with and without cross terms, we attempt to evaluate the importance of cross diffusion.

We concentrate on the local acceleration and loss processes in the center ($L = 4.5$) of the outer radiation belt. The computational range covers $\alpha_e \in [0°, 90°]$ and $E_k \in [0.2\,\text{MeV}, 10.0\,\text{MeV}]$. The uniform grids are adopted in the (α_e, ξ) space with grid numbers 91×81. The time-step is set to be $\Delta t = 1$ s. It should be noted that only the electron fluxes in the energy range $E_k \in [0.2\,\text{MeV}, 5.0\,\text{MeV}]$ are plotted to clearly identify the ~MeV electron evolution.

The initial electron PSD is assumed to obey the Kappa-type distribuion (Vasyliunas 1968; Maksimovic et al. 1997a, b; Viñas et al. 2005; Xiao 2006; Xiao et al. 2008a, b, c)

$$f(t = 0, \alpha_e, p) = C \left(\frac{p \sin \alpha_e}{\theta_\kappa} \right)^{2l} \left[1 + \frac{p^2}{\kappa \theta_\kappa^2} \right]^{-(\kappa + l + 1)}, \tag{2.46}$$

$$C = \frac{N \Gamma(\kappa + l + 1)}{\pi^{3/2} \theta_\kappa^{3/2} \kappa^{(l+3/2)} \Gamma(l + 1) \Gamma(\kappa - 1/2)}, \tag{2.47}$$

with the loss cone prameter l, effective thermal parameter θ_κ^2 electron density N, spectral index κ, and Gamma function Γ. Here these parameters are empirically chosen to be $\theta_\kappa^2 = 0.15$, $l = 0.5$ and $\kappa = 6$ (Xiao et al. 2009).

At the loss cone boundary $\alpha_e = \alpha_L$, the electron PSD is set to be zero. At $\alpha_e = 90°$, the equivalent extrapolation boundary condition is adopted. At the upper and lower energy boundaries, the electron PSDs are fixed.

2.3.1 Chorus

Based on the previous works (Meredith et al. 2001, 2002, 2003; Horne et al. 2005; Li et al. 2007), we choose different parameters for nightside and dayside chorus waves and the corresponding background plasma. We assume the constant electron density along the geomagnetic file lines. The equatorial ratio between electron plasma frequency ω_{pe} and gyrofrequency $|\Omega_e|$ is set to be 3.8 in the nightside and 4.6 in the dayside. At the nightside, the chorus waves are distributed over the latitudinal region $|\lambda| < 15°$ with the spectral parameters $B_t = 50$ pT, $\omega_1 = 0.05|\Omega_e|$, $\omega_2 = 0.65|\Omega_e| \delta\omega = 0.15|\Omega_e|$, $\omega_m = 0.35|\Omega_e|$, $X_m = 0$, $X_\omega = 0.577$, $X_1 = 0$ and

$X_2 = 1$. At the dayside, the chorus waves are distributed in the latitudinal region $\lambda \leq 35^\circ$ with the spectral parameters $B_t = 10^{0.75+0.04\lambda}$ pT, $\omega_1 = 0.1|\Omega_e|$, $\omega_2 = 0.3|\Omega_e|$, $\delta\omega = 0.1|\Omega_e|$, $\omega_m = 0.2|\Omega_e|$, $X_m = 0$, $X_\omega = 0.577$, $X_1 = 0$ and $X_2 = 1$. On both sides, the resonance orders $n = 0, \pm 1, \ldots, \pm 5$ are taken into account. It should be noted that these parameters have been frequently used to simulate the chorus-driven radiation belt electron evolution during storms (e.g., Li et al. 2007; Shprits et al. 2009a, b; Subbotin et al. 2010; Su et al. 2009a, 2011a, b, c, 2010a, c; Xiao et al. 2009, 2010b).

Figure 2.1 shows the two-dimensional distribution of diffusion rates for the nightside and dayside chorus waves in the (α_e, E_k) space, and Fig. 2.2 plots the diffusion coefficient profiles at the selected energies. Obviously, the pitch-angle and momentum diffusion rates behave smoothly, but the cross diffusion rates exhibit rapid and significant fluctuations which can easily lead to numerical instability (Albert 2004; Albert and Young 2005). The maximum values of pitch-angle diffusion rates are about 10 times larger than those of momentum diffusion rates, and the maximum values of momentum and cross diffusion rates are generally comparable (implying that the ignoring of cross diffusion rates is unreasonable). The diffusion rates for nightside and dayside chorus waves have quite different distributions. The diffusion rates of nightside chorus waves peak at the large equatorial pitch-angles (near $\alpha_e = 90^\circ$). In contrast, the diffusion rates of dayside chorus waves peak at the small equatorial pitch-angles (near $\alpha_e = \alpha_L$).

The energetic electrons drift around the Earth on closed paths, quasi-periodically passing through the spatial regions with different types of waves. We assume that the nightside and dayside chorus waves are distributed 25 % of drift paths, respectively. The drift-averaged diffusion coefficients are input into Eq. (2.9) to simulate the evolution of electron PSD f and flux $j = p^2 f$. Figures 2.3 and 2.4 present the electron flux evolution with and without cross diffusion. As shown in the simulations with cross diffusion, the chorus waves can effectively accelerate the electrons and enhance significantly the energetic electron fluxes especially at the large equatorial pitch-angles ($\alpha_e > 45^\circ$). Within two days, the equatorially-trapped electron fluxes at energies $E_k = 0.5$, 1.0 and 2.0 MeV can increase by about 3, 30, and 60 times, respectively. The ignoring of cross diffusion can largely overestimate the acceleration effect of chorus waves. After two days, the equatorially-trapped electron fluxes at energies $E_k = 0.5$, 1.0, and 2.0 MeV are overestimated by about 3, 6, and 5 times at the large pitch-angles, and by about 10, 50, and 200 times at the small pitch-angles, respectively. These results suggest that cross diffusion plays an important role in chorus-electron cyclotron resonance.

2.3.2 Hiss

We assume constant electron density along the magnetic field line, and set the equatorial $\omega_{pe}/|\Omega_e| = 15$ (Li et al. 2007). The hiss waves are distributed in the latitudinal region $\lambda < 40^\circ$ with spectral properties $B_t = 0.1$ nT, $\omega_1 = 0.01|\Omega_e|$, $\omega_2 = 0.21|\Omega_e|$,

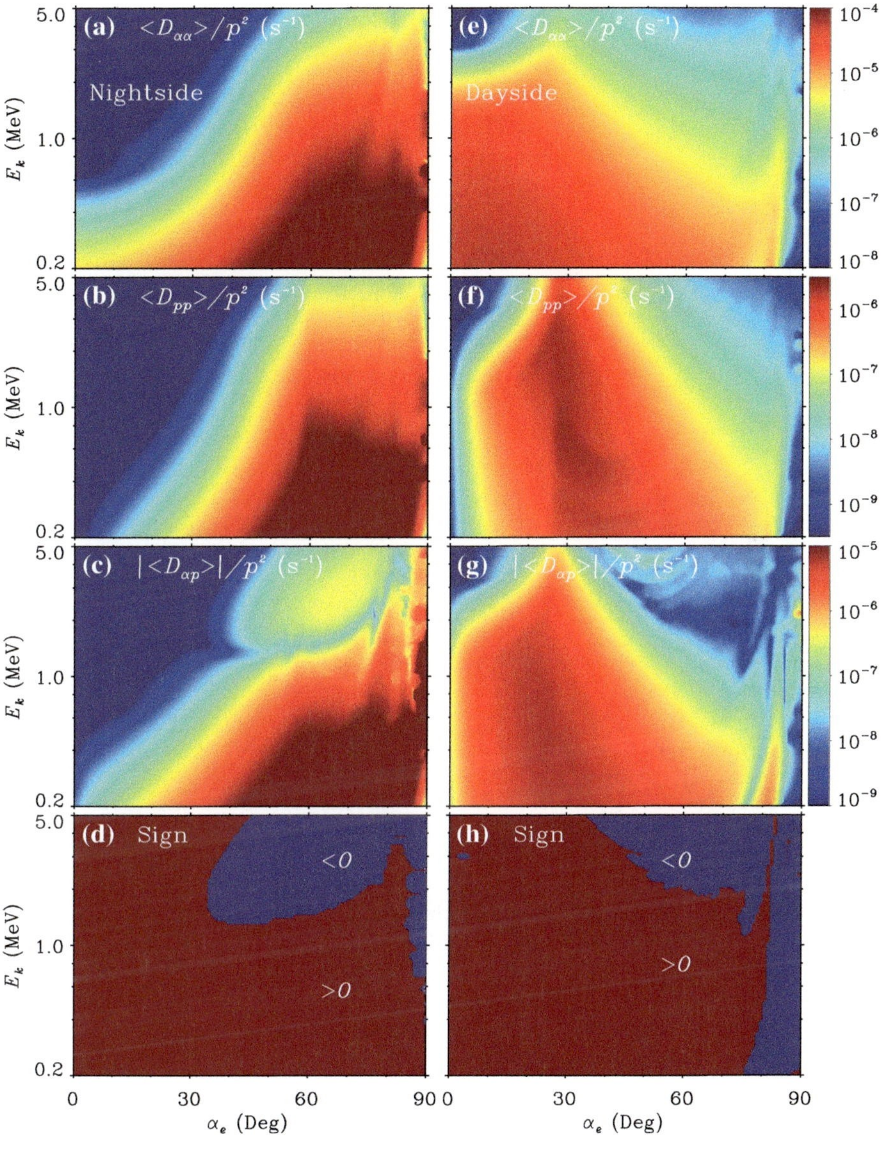

Fig. 2.1 Distribution of nightside (*left*) and dayside (*right*) chorus-driven diffusion rates in the (α_e, E_k) space

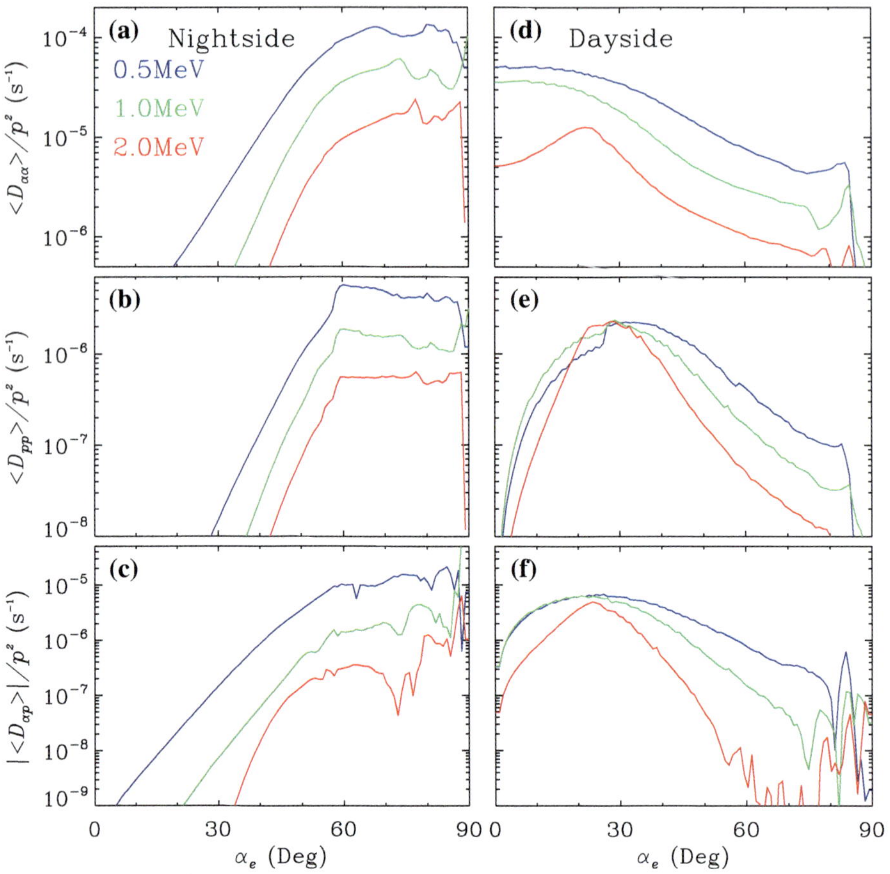

Fig. 2.2 Nightside (*left*) and dayside (*right*) chorus-driven diffusion rate profiles at energies $E_k = 0.5$, 1.0 and 2.0 MeV

$\delta\omega = 0.03|\Omega_e|$, $\omega_m = 0.06|\Omega_e|$, $X_m = 0$, $X_\omega = 0.577$, $X_1 = 0$ and $X_2 = 1$. Similar to chorus waves, the cyclotron resonance orders $n = 0, \pm 1, \ldots, \pm 5$ are taken into account for hiss waves. These parameters have been frequently adopted to simulate the hiss-driven radiation belt electron evolution (e.g., Li et al. 2007; Shprits et al. 2009b; Xiao et al. 2009; Su et al. 2010a, 2011a, b, c).

The distributions and profiles of obtained hiss-driven diffusion rates are shown in Figs. 2.5 and 2.6. The maximum diffusion rates occur at the large equatorial pitch-angles and small energies. As the energy increases, the diffusion rates decrease gradually. The maximum values of pitch-angle and cross diffusion rates are 3000 and 50 times larger than the momentum diffusion rates, respectively.

We assume that the hiss waves are distributed in the 15 % of electron drift pathes (Li et al. 2007), input the drift-averaged diffusion coefficients into the Eq. (2.9), and obtain the electron flux $j = p^2 f$ evolution (given in Figs. 2.7 and 2.8). As shown

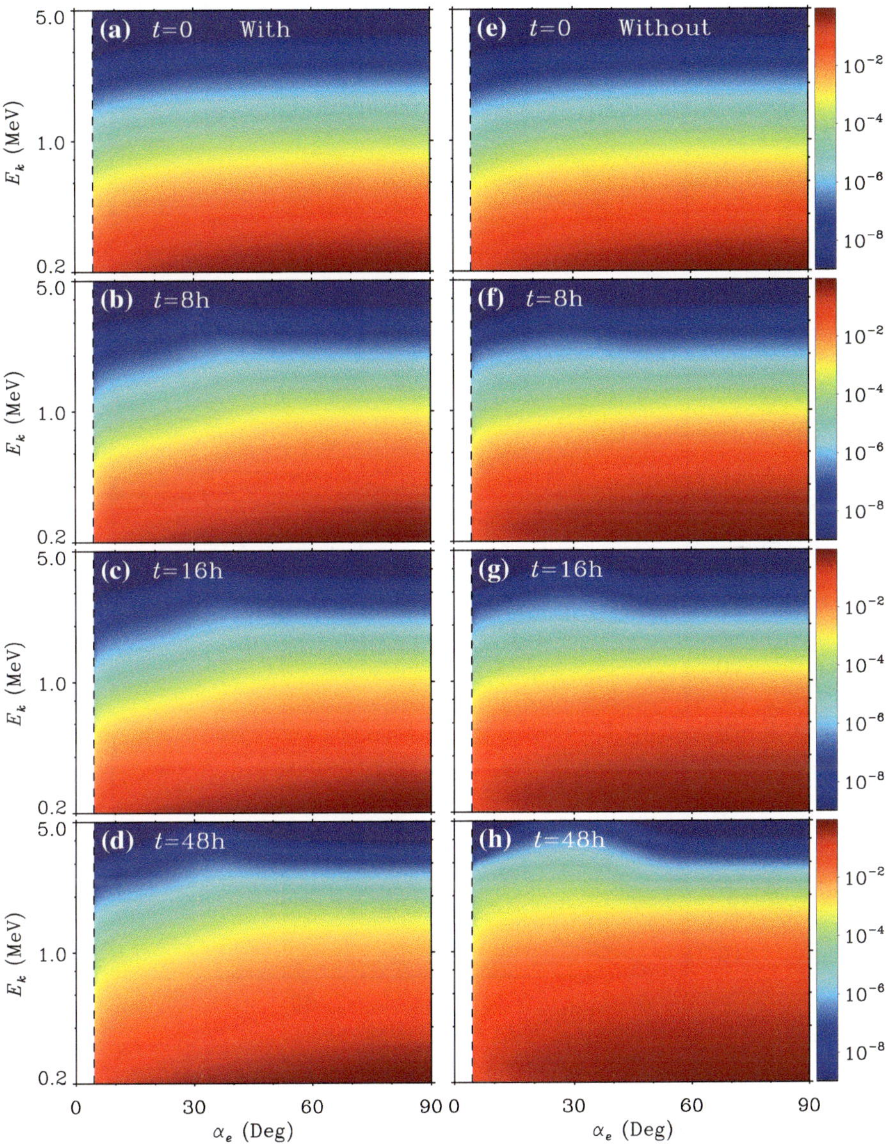

Fig. 2.3 Evolution of chorus-driven electron flux $j = p^2 f$ (arbitrary unit) in the (α_e, E_k) space. The *left* and *right* panels correspond to the simulations with and without cross diffusion

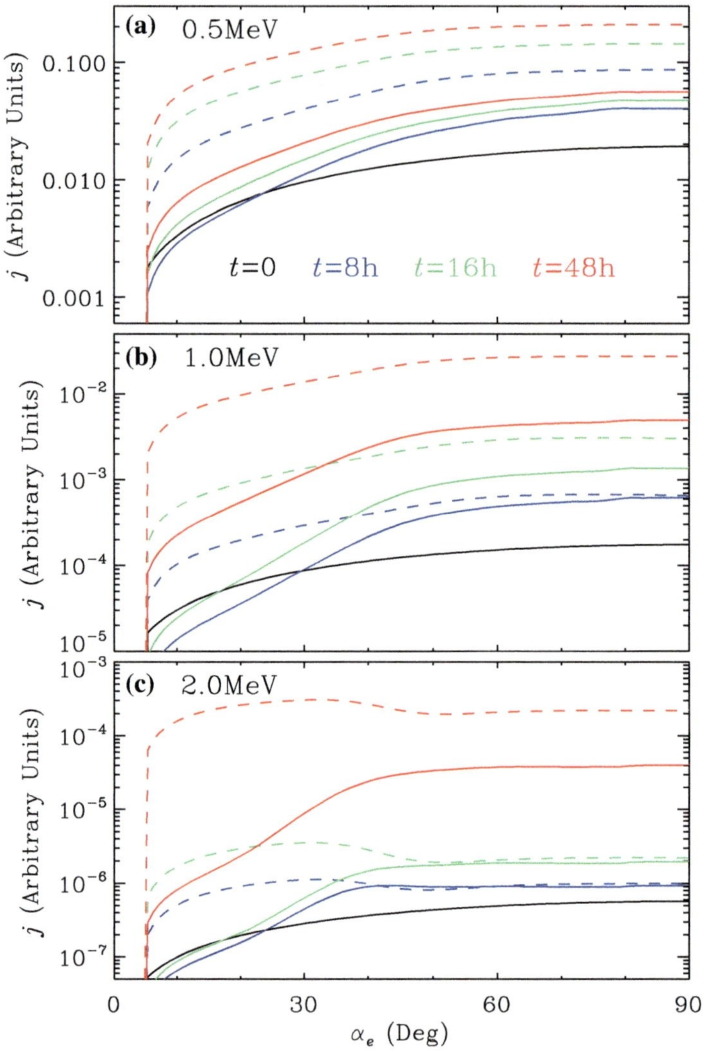

Fig. 2.4 Electron flux profiles $j = p^2 f$ at the selected time points. The *solid* and *dashed lines* represent the simulations with and without cross diffusion

in the simulations with cross diffusion, the hiss waves can effectively scatter the energetic electrons especially at the large pitch-angles, drive them toward the loss cone, and produce the electron precipitation loss. Since the diffusion coefficients generally decrease with increasing energy, the hiss waves show weaker loss effect on the electrons with larger energies. Within two days, the equatorially-trapped electron fluxes at energies $E_k = 0.5$, 1.0, and 2.0 MeV are cut to 1/5, 1/2, and 2/3, respectively.

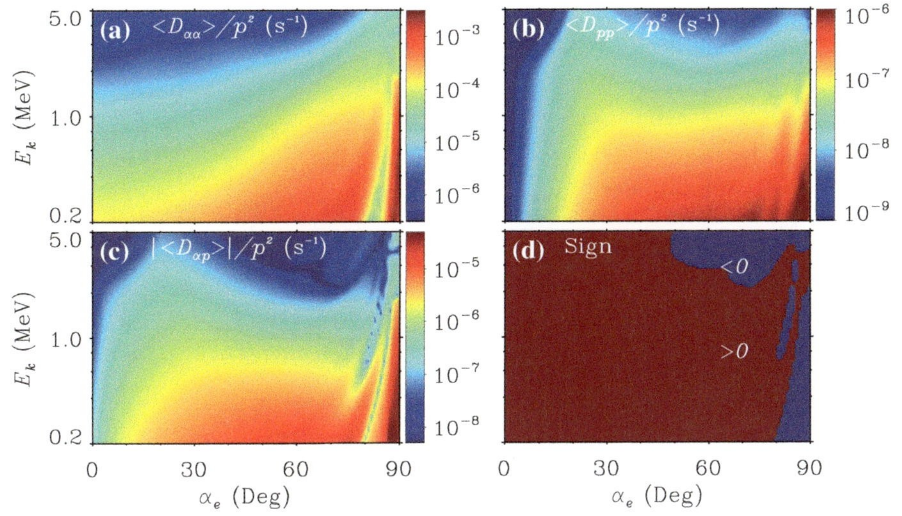

Fig. 2.5 Distribution of hiss-driven diffusion rates in the (α_e, E_k) space

Ignoring of cross diffusion can lead to the overestimate of electron fluxes by <2 times, indicating that cross diffusion plays an insignificant role in hiss-electron cyclotron resonance.

2.3.3 EMIC

The electron density is fixed along the magnetic field line, and the equatorial $\omega_{pe}/|\Omega_e|$ is specified as 15 (Li et al. 2007). It should be mentioned that, the statistical characteristics of EMIC waves are not clear (Shprits et al. 2009a). Based on previous works (Li et al. 2007; Summers et al. 2007b), we concentrate here on the parallel-propagating He^+ band EMIC waves. The background ion composition is assumed to be $70\,\%H^+ + 20\,\%He^+ + 10\,\%O^+$. The EMIC waves are distributed in the low latitude region $\lambda < 15°$ with spectral parameters $B_t = 1.0$ nT, $\omega_1 = 3.45\Omega_{O^+}$, $\omega_2 = 3.95\Omega_{O^+}$, $\delta\omega = 0.25\Omega_{O^+}$ and $\omega_m = 3.70\Omega_{O^+}$. For such parallel-propagating L-mode waves, only the $n = 1$ order cyclotron resonance can occur (Summers 2005).

The obtained diffusion rate distributions and profiles are plotted in Figs. 2.9 and 2.10. The modeled EMIC waves can only resonate with the high energy electrons (Summers 2005; Summers et al. 2007a). The minimum resonant energy for the present EMIC waves is about $E_{kmin} = 0.4$ MeV. As the energy increases, the resonant pitch-angles extend toward $\alpha_e = 90°$. The maximum values of pitch-angle diffusion rates are about 2×10^7 and 4×10^3 times larger than those of momentum and cross diffusion rates, respectively, indicating the dominance of pitch-angle diffusion.

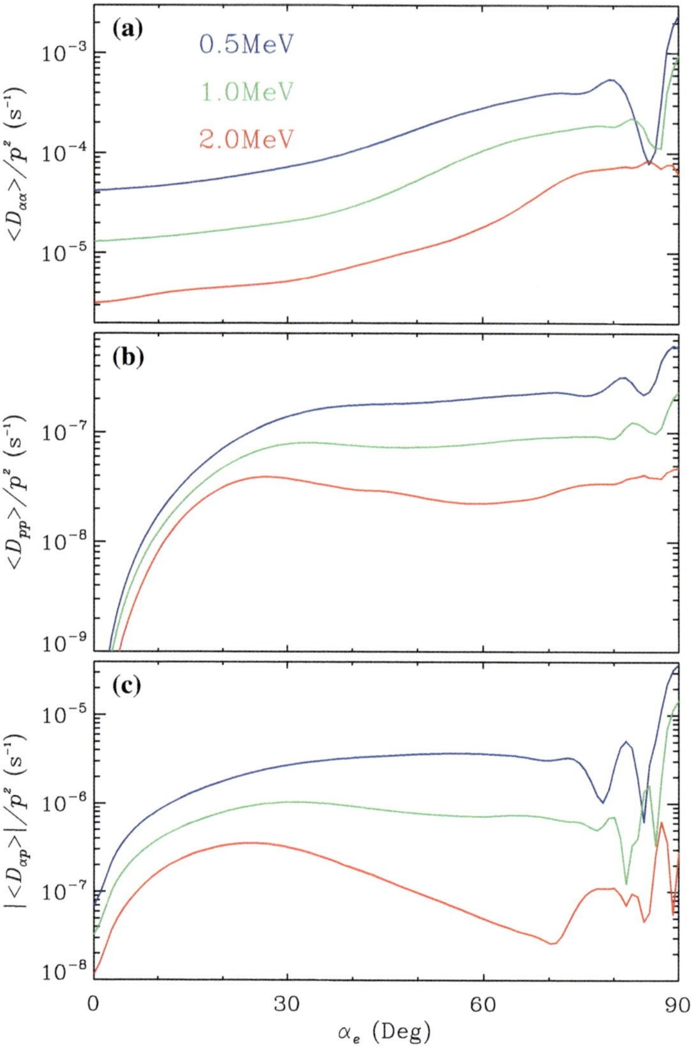

Fig. 2.6 Hiss-driven diffusion rate profiles at energies $E_k = 0.5$, 1.0 and 2.0 MeV

The EMIC waves are assumed to be distributed in the 5 % of drift pathes (Li et al. 2007; Shprits et al. 2009b; Su et al. 2010a, 2011a, b, c), and the obtained electron flux evolution is plotted in Figs. 2.11 and 2.12. Due to the dominance of pitch-angle diffusion, there is no notable difference between the simulation results including and ignoring cross diffusion. The EMIC waves can cause rapid precipitation loss of energetic electrons. After 12 h, the electron fluxes near the loss cone at energies $E_k = 0.5$, 1.0 and 2.0 MeV are cut to 1/50, 1/200, and 1/3000, respectively. The EMIC waves cannot scatter the energetic electron near $\alpha_e = 90°$, but the hiss waves

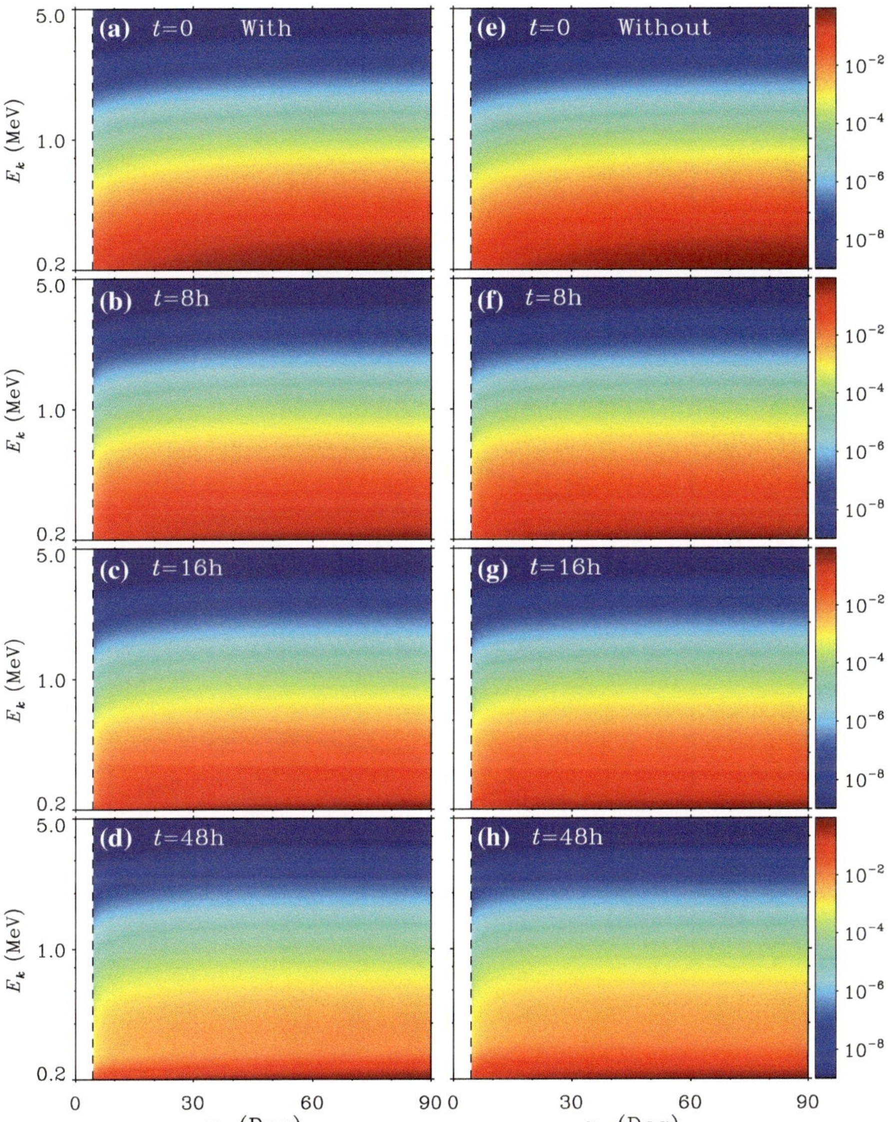

Fig. 2.7 Evolution of chorus-driven electron flux $j = p^2 f$ (arbitrary unit) in the (α_e, E_k) space. The *left* and *right panels* correspond to the simulations with and without cross diffusion

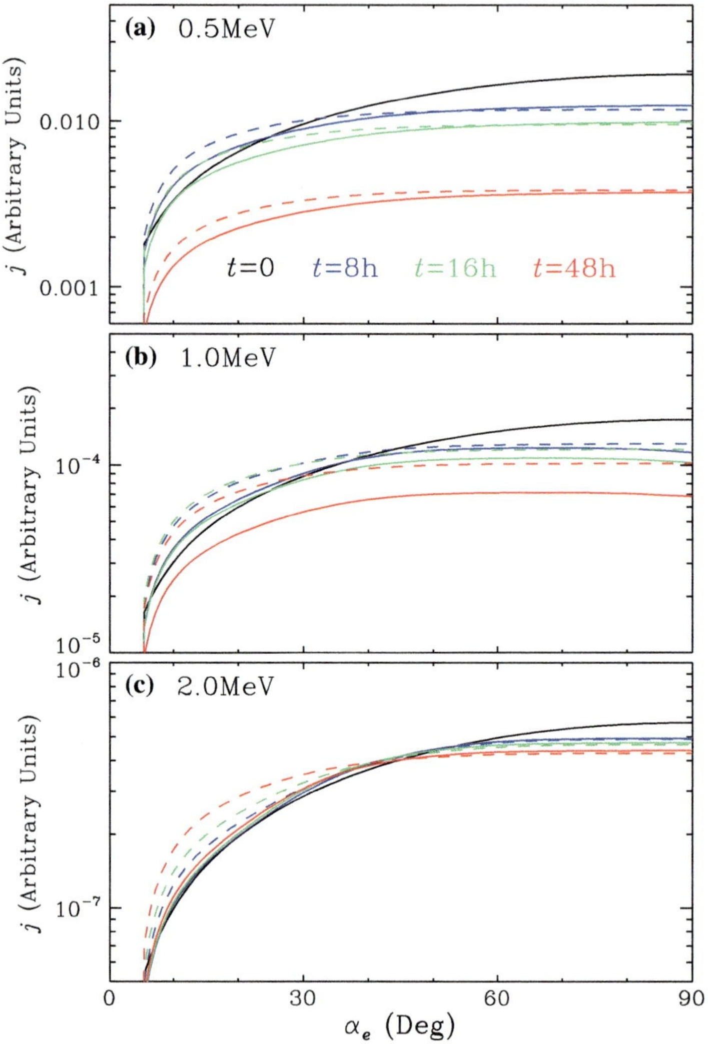

Fig. 2.8 Electron flux profiles $j = p^2 f$ at the selected time points. The *solid* and *dashed lines* represent the simulations with and without cross diffusion

can drive the electrons with the large pitch-angles toward the loss cone. Hence, the combination of hiss and EMIC waves would produce more effective precipitation loss of radiation belt electrons (Li et al. 2007; Su et al. 2011c).

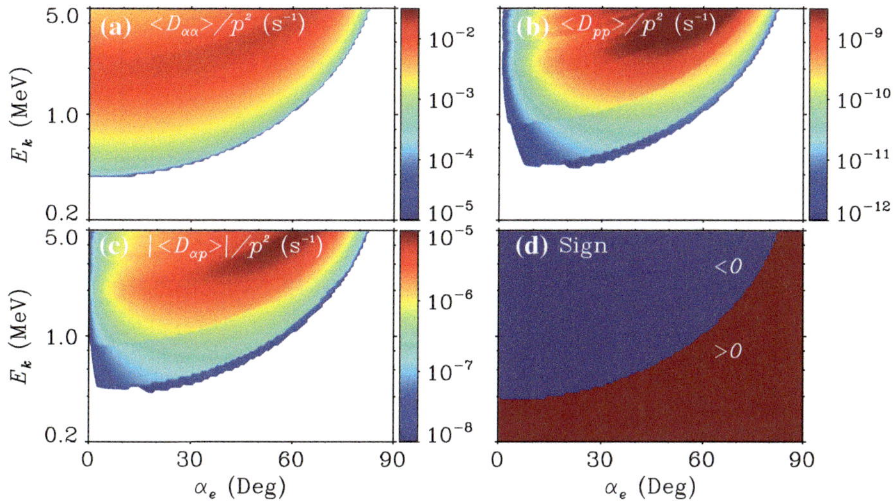

Fig. 2.9 Distribution of EMIC-driven diffusion rates in the (α_e, E_k) space

2.4 Conclusions and Discussions

We have developed an electron radiation belt local diffusion model based on the quasilinear theory. The quasilinear diffusion equation involves the diffusion in pitch-angle, momentum, and cross terms. The including of cross diffusion often yields numerical instability for the common finite difference methods (Albert 2004; Albert and Young 2005). We propose an efficient, stable, and easily-programming numerical scheme, named HFD, to fully solve this diffusion equation (Su et al. 2009c, 2011c; Xiao et al. 2009). This numerical scheme has been widely used by our research group (Su et al. 2009a, b, c, 2010a, b, c, 2011a, b, c; Xiao et al. 2009, 2010b, a, 2011) and other researchers (Fok et al. 2010; Thorne et al. 2010; Tao et al. 2011b; Zheng et al. 2011).

Using this local diffusion model, we have analyzed the effect of chorus, hiss and EMIC on the radiation belt electrons, and particularly identify the importance of cross diffusion.

1. The present simulation results with cross diffusion support the previous finding. The chorus waves can effectively accelerate the energetic electrons, especially at the large pitch-angles: the equatorially trapped energetic electron fluxes can increase by about 1–2 orders of magnitude within 2 days. The hiss waves can drive the electrons at the large pitch-angles toward the loss cone: the equatorially-trapped energetic electron fluxes can be cut to 1/5–2/3 within 2 days. The EMIC waves can effectively scatter the relativistic electrons into the loss cone: the relativistic electron fluxes near the loss cone can be cut to 1/3000–1/50 within 12 h.

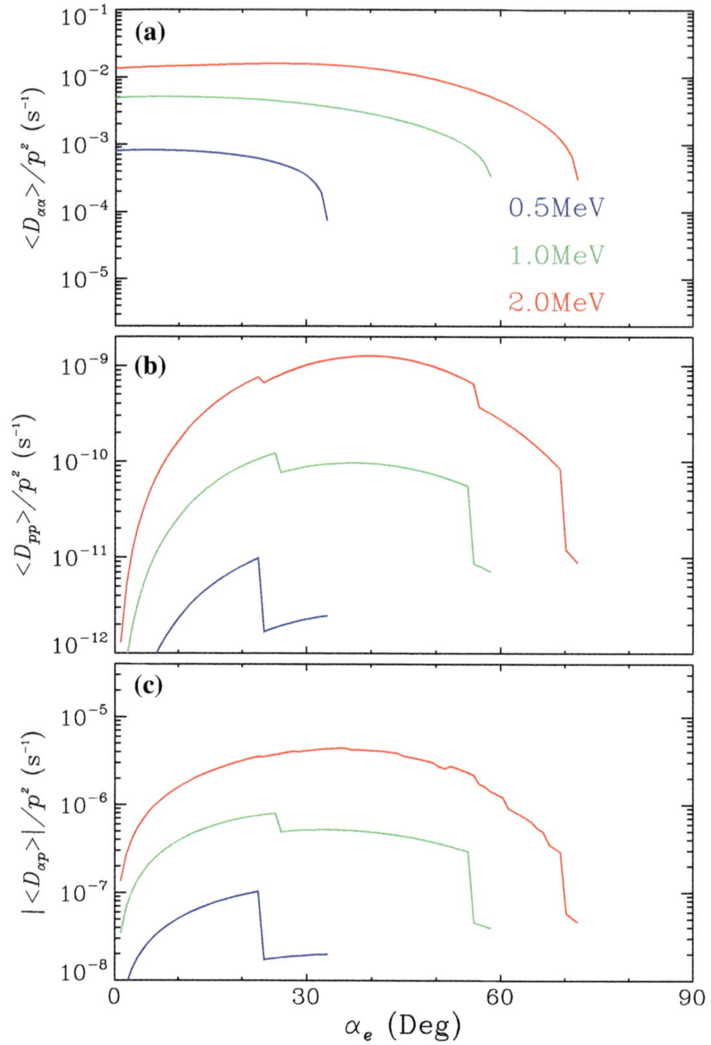

Fig. 2.10 EMIC-driven diffusion rate profiles at energies $E_k = 0.5$, 1.0 and 2.0 MeV

2. Cross diffusion plays different roles in the different wave-particle interactions. Ignoring of cross diffusion can significantly overestimate the electron fluxes for cyclotron resonance with chorus, slightly overestimate the electron fluxes for the cyclotron resonance with hiss, but have no effect on the electron fluxes for cyclotron resonance with EMIC.

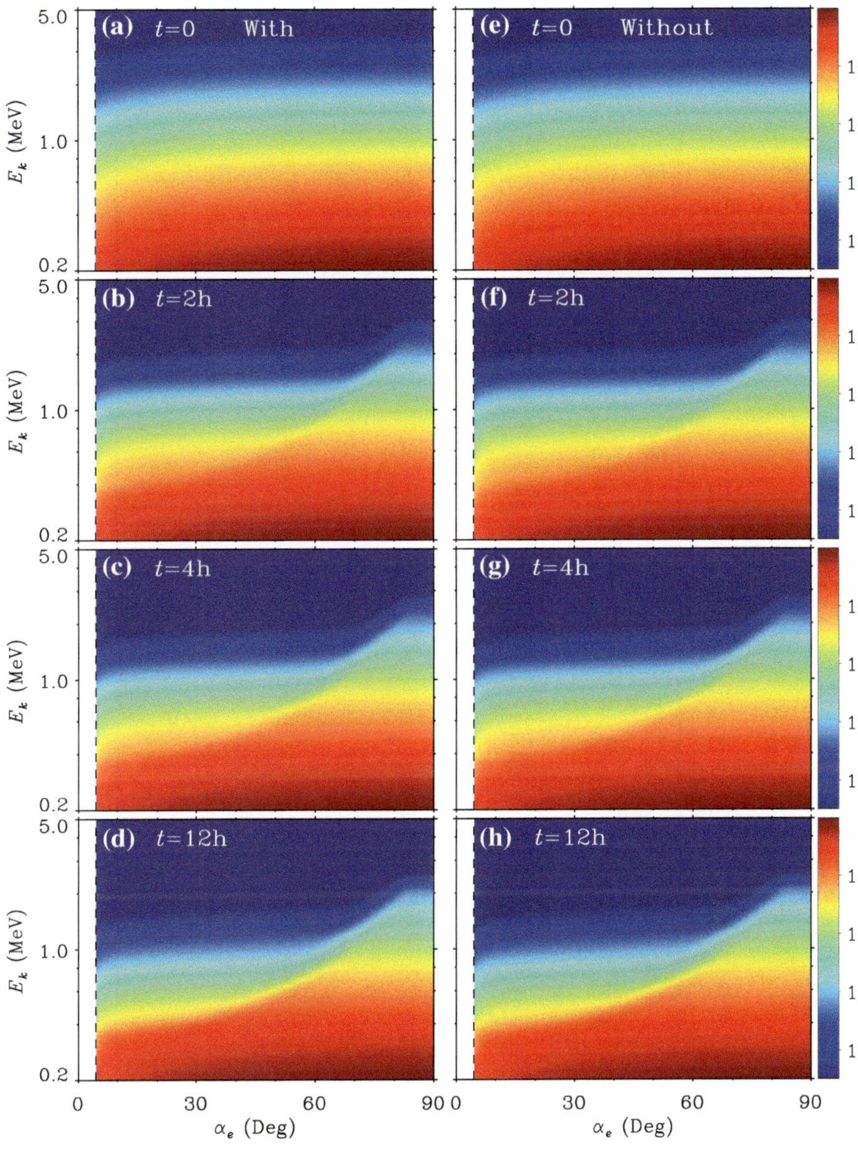

Fig. 2.11 Evolution of EMIC-driven electron flux $j = p^2 f$ (arbitrary unit) in the (α_e, E_k) space. The *left* and *right panels* correspond to the simulations with and without cross diffusion

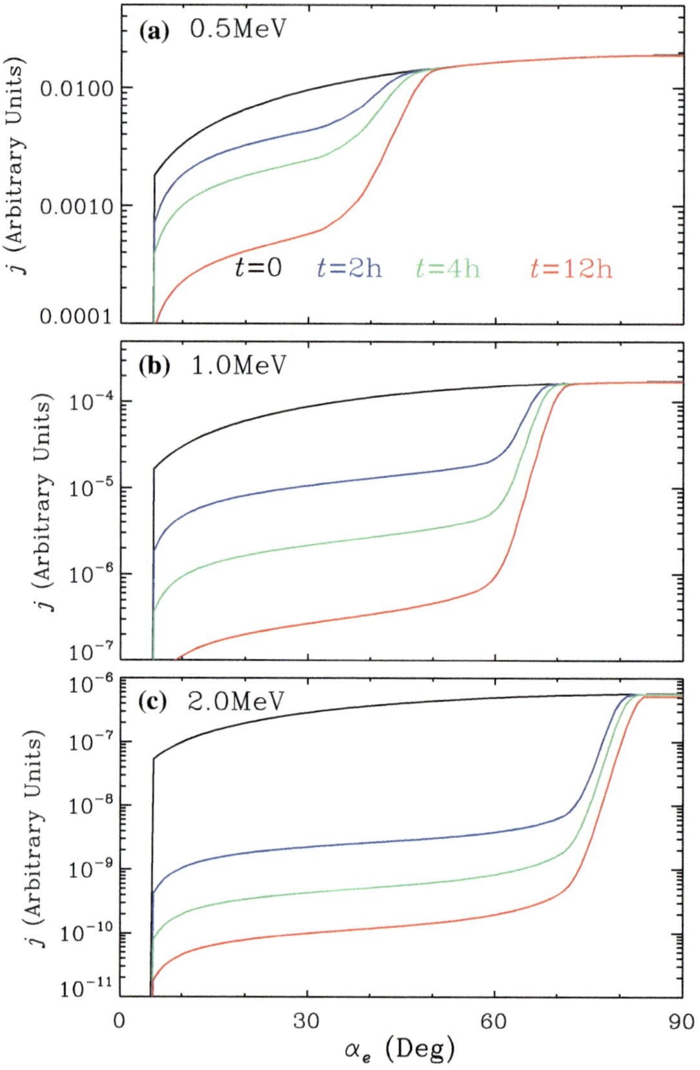

Fig. 2.12 Electron flux profiles $j = p^2 f$ at the selected time points. The *solid* and *dashed lines* represent the simulations with and without cross diffusion, which coincide with each other

References

Albert JM (1999) Analysis of quasi-linear diffusion coefficients. J Geophys Res 104:2429–2442. doi:10.1029/1998JA900113

Albert JM (2002) Nonlinear interaction of outer zone electrons with VLF waves. Geophys Res Lett 29:1275. doi:10.1029/2001GL013941

Albert JM (2003) Evaluation of quasi-linear diffusion coefficients for EMIC waves in a multispecies plasma. J Geophys Res 108(A8):1249. doi:10.1029/2002JA009792

Albert JM (2004) Using quasi-linear diffusion to model acceleration and loss from wave-particle interactions. Space Weather 2:9. doi:10.1029/2004SW000069

Albert JM (2005) Evaluation of quasi-linear diffusion coefficients for whistler mode waves in a plasma with arbitrary density ratio. J Geophys Res 110(A9):A03,218, doi:10.1029/2004JA010844

Albert JM (2008) Efficient approximations of quasi-linear diffusion coefficients in the radiation belts. J Geophys Res 113(A06):208. doi:10.1029/2007JA012936

Albert JM, Bortnik J (2009) Nonlinear interaction of radiation belt electrons with electromagnetic ion cyclotron waves. Geophys Res Lett 36(L12):110. doi:10.1029/2009GL038904

Albert JM, Young SL (2005) Multidimensional quasi-linear diffusion of radiation belt electrons. Geophys Res Lett 32(L14):110. doi:10.1029/2005GL023191

Albert JM, Meredith NP, Horne RB (2009) Three-dimensional diffusion simulation of outer radiation belt electrons during the october 9, 1990, magnetic storm. J Geophys Res 114(A09):214. doi: 10.1029/2009JA014336

Beutier T, Boscher D, France M (1995) Salammbo: a three-dimensional simulation of the proton radiation belt. J Geophys Res 100:17181–17188. doi:10.1029/94JA02728

Bortnik J, Thorne RM (2010) Transit time scattering of energetic electrons due to equatorially confined magnetosonic waves. J Geophys Res 115(A07):213. doi:10.1029/2010JA015283

Bortnik J, Thorne RM, Inan US (2008a) Nonlinear interaction of energetic electrons with large amplitude chorus. Geophys Res Lett 35(L21):102. doi:10.1029/2008GL035500

Bortnik J, Thorne RM, Meredith NP (2008b) The unexpected origin of plasmaspheric hiss from discrete chorus emissions. Nature 452:62–66. doi:10.1038/nature06741

Bortnik J, Li W, Thorne RM, Angelopoulos V, Cully C, Bonnell J, Le Contel O, Roux A (2009) An observation linking the origin of plasmaspheric hiss to discrete chorus emissions. Science 324:775. doi:10.1126/science.1171273

Bourdarie S, Boscher D, Beutier T, Sauvaud J, Blanc M (1997) Electron and proton radiation belt dynamic simulations during storm periods: a new asymmetric convection-diffusion model. J Geophys Res 102:17541–17552. doi:10.1029/97JA01305

Burtis WJ, Helliwell RA (1975) Magnetospheric chorus—amplitude and growth rate. J Geophys Res 80:3265–3270. doi:10.1029/JA080i022p03265

Cattell C, Wygant JR, Goetz K, Kersten K, Kellogg PJ, von Rosenvinge T, Bale SD, Roth I, Temerin M, Hudson MK, Mewaldt RA, Wiedenbeck M, Maksimovic M, Ergun R, Acuna M, Russell CT (2008) Discovery of very large amplitude whistler-mode waves in Earth's radiation belts. Geophys Res Lett 35(L01):105. doi:10.1029/2007GL032009

Cully CM, Bonnell JW, Ergun RE (2008) THEMIS observations of long-lived regions of large-amplitude whistler waves in the inner magnetosphere. Geophys Res Lett 35:L17S16, doi: 10.1029/2008GL033643

Fok MC, Glocer A, Zheng Q, Horne RB, Meredith NP, Albert JM, Nagai T (2010) Recent developments in the radiation belt environment model. J Atmos Sol Terr Phys. 73:1435–1443. doi: 10.1016/j.jastp.2010.09.033

Fraser BJ, Singer HJ, Hughes WJ, Wygant JR, Anderson RR, Hu YD (1996) Crres poynting vector observations of electromagnetic ion cyclotron waves near the plasmapause. J Geophys Res 101:15331–15344. doi:10.1029/95JA03480

Glauert SA, Horne RB (2005) Calculation of pitch angle and energy diffusion coefficients with the PADIE code. J Geophys Res 110(A04):206. doi:10.1029/2004JA010851

Horne RB, Thorne RM (1998) Potential waves for relativistic electron scattering and stochastic acceleration during magnetic storms. Geophys Res Lett 25:3011–3014. doi:10.1029/98GL01002
Horne RB, Thorne RM (2003) Relativistic electron acceleration and precipitation during resonant interactions with whistler-mode chorus. Geophys Res Lett 30:1527. doi:10.1029/2003GL016973
Horne RB, Glauert SA, Thorne RM (2003a) Resonant diffusion of radiation belt electrons by whistler-mode chorus. Geophys Res Lett 30:1493. doi:10.1029/2003GL016963
Horne RB, Meredith NP, Thorne RM, Heynderickx D, Iles RHA, Anderson RR (2003b) Evolution of energetic electron pitch angle distributions during storm time electron acceleration to megaelectronvolt energies. J Geophys Res 108:1016. doi:10.1029/2001JA009165
Horne RB, Thorne RM, Glauert SA, Albert JM, Meredith NP, Anderson RR (2005) Timescale for radiation belt electron acceleration by whistler mode chorus waves. J Geophys Res 110(A03):225. doi:10.1029/2004JA010811
Kennel CF, Engelmann F (1966) Velocity space diffusion from weak plasma turbulence in a magnetic field. Phys Fluids 9:2377–2388. doi:10.1063/1.1761629
Kim J, Ryu D, Jones TW, Hong SS (1999) A multidimensional code for isothermal magnetohydrodynamic flows in astrophysics. Astrophys J 514:506–519. doi:10.1086/306915
Koons HC, Roeder JL (1990) A survey of equatorial magnetospheric wave activity between 5 and 8 R_E. Planet Space Sci 38:1335–1341. doi:10.1016/0032-0633(90)90136-E
Kozyra JU, Rasmussen CE, Miller RH, Lyons LR (1994) Interaction of ring current and radiation belt protons with ducted plasmaspheric hiss. 1: diffusion coefficients and timescales. J Geophys Res 99:4069–4084. doi:10.1029/93JA01532
Li L, Cao J, Zhou G (2005) Combined acceleration of electrons by whistler-mode and compressional ULF turbulences near the geosynchronous orbit. J Geophys Res 110(A03):203. doi:10.1029/2004JA010628
Li W, Shprits YY, Thorne RM (2007) Dynamic evolution of energetic outer zone electrons due to wave-particle interactions during storms. J Geophys Res 112(A11):A10,220, doi:10.1029/2007JA012368
Lyons LR (1974a) Electron diffusion driven by magnetospheric electrostatic waves. J Geophys Res 79:575–580. doi:10.1029/JA079i004p00575
Lyons LR (1974b) General relations for resonant particle diffusion in pitch angle and energy. J Plasma Phys 12:45. doi:10.1017/S0022377800024910
Lyons LR (1974c) Pitch angle and energy diffusion coefficients from resonant interactions with ion-cyclotron and whistler waves. J Plasma Phys 12:417–432
Lyons LR, Thorne RM (1973) Equilibrium structure of radiation belt electrons. J Geophys Res 78:2142–2149. doi:10.1029/JA078i013p02142
Lyons LR, Williams DJ (1984) Quantitative aspects of magnetospheric physics. Springer, New York
Maksimovic M, Pierrard V, Lemaire JF (1997a) A kinetic model of the solar wind with Kappa distribution functions in the corona. Astron Astrophys 324:725–734
Maksimovic M, Pierrard V, Riley P (1997b) Ulysses electron distributions fitted with Kappa functions. Geophys Res Lett 24:1151–1154. doi:10.1029/97GL00992
Meredith NP, Horne RB, Anderson RR (2001) Substorm dependence of chorus amplitudes: implications for the acceleration of electrons to relativistic energies. J Geophys Res 106:13165–13178. doi:10.1029/2000JA900156
Meredith NP, Horne RB, Iles RHA, Thorne RM, Heynderickx D, Anderson RR (2002) Outer zone relativistic electron acceleration associated with substorm-enhanced whistler mode chorus. J Geophys Res 107:1144. doi:10.1029/2001JA900146
Meredith NP, Horne RB, Thorne RM, Anderson RR (2003) Favored regions for chorus-driven electron acceleration to relativistic energies in the earth's outer radiation belt. Geophys Res Lett 30:1871. doi:10.1029/2003GL017698
Meredith NP, Horne RB, Thorne RM, Summers D, Anderson RR (2004) Substorm dependence of plasmaspheric hiss. J Geophys Res 109(A06):209. doi:10.1029/2004JA010387

Meredith NP, Horne RB, Glauert SA, Anderson RR (2007) Slot region electron loss timescales due to plasmaspheric hiss and lightning-generated whistlers. J Geophys Res 112:8214. doi:10.1029/2007JA012413

Nunn D (1974) A self-consistent theory of triggered VLF emissions. Planet Space Sci 22:349–378. doi:10.1016/0032-0633(74)90070-1

O'Brien TP, Lorentzen KR, Mann IR, Meredith NP, Blake JB, Fennell JF, Looper MD, Milling DK, Anderson RR (2003) Energization of relativistic electrons in the presence of ULF power and MeV microbursts: evidence for dual ULF and VLF acceleration. J Geophys Res 108:1329. doi:10.1029/2002JA009784

Orlova KG, Shprits YY (2010) Dependence of pitch-angle scattering rates and loss timescales on the magnetic field model. Geophys Res Lett 370(L05):105. doi:10.1029/2009GL041639

Parrot M, Gaye CA (1994) A statistical survey of ELF waves in a geostationary orbit. Geophys Res Lett 21:2463–2466. doi:10.1029/94GL01700

Roth I, Temerin M, Hudson MK (1999) Resonant enhancement of relativistic electron fluxes during geomagnetically active periods. Ann Geophys 17:631–638. doi:10.1007/s00585-999-0631-2

Russell CT, Holzer RE, Smith EJ (1969) OGO 3 observations of ELF noise in the magnetosphere. 1. Spatial extent and frequency of occurrence. J Geophys Res 74:755–777. doi: 10.1029/JA074i003p00755

Santolík O, Gurnett DA, Pickett JS, Parrot M, Cornilleau-Wehrlin N (2003) Spatio-temporal structure of storm-time chorus. J Geophys Res 108:1278. doi:10.1029/2002JA009791

Santolík O, Gurnett DA, Pickett JS, Parrot M, Cornilleau-Wehrlin N (2004) A microscopic and nanoscopic view of storm-time chorus on 31 March 2001. Geophys Res Lett 31(L02):801. doi: 10.1029/2003GL018757

Schulz M, Lanzerotti LJ (1974) Particle diffusion in the radiation belts, in physics and chemistry in space, vol 7. Springer, New York

Shprits YY, Thorne RM, Horne RB, Glauert SA, Cartwright M, Russell CT, Baker DN, Kanekal SG (2006) Acceleration mechanism responsible for the formation of the new radiation belt during the 2003 halloween solar storm. Geophys Res Lett 33(L05):104. doi:10.1029/2005GL024256

Shprits YY, Elkington SR, Meredith NP, Subbotin DA (2008) Review of modeling of losses and sources of relativistic electrons in the outer radiation belt I: radial transport. J Atmos Sol Terr Phys 70:1679–1693. doi:10.1016/j.jastp.2008.06.008

Shprits YY, Subbotin DA, Meredith NP, Elkington SR (2008) Review of modeling of losses and sources of relativistic electrons in the outer radiation belt II: local acceleration and loss. J Atmos Sol Terr Phys 70:1694–1713. doi:10.1016/j.jastp.2008.06.014

Shprits YY, Chen L, Thorne RM (2009a) Simulations of pitch angle scattering of relativistic electrons with MLT-dependent diffusion coefficients. J Geophys Res 114(A03):219. doi:10.1029/2008JA013695

Shprits YY, Subbotin D, Ni B (2009b) Evolution of electron fluxes in the outer radiation belt computed with the VERB code. J Geophys Res 114(A11):209. doi:10.1029/2008JA013784

Smith EJ, Frandsen AMA, Tsurutani BT, Thorne RM, Chan KW (1974) Plasmaspheric hiss intensity variations during magnetic storms. J Geophys Res 79:2507–2510. doi:10.1029/JA079i016p02507

Stix TH (1992) Waves in plasmas. American Institute of Physics, New York

Strang G (1968) On the construction and comparison of difference schemes. SIAM J Numer Anal 5:506

Su Z, Zheng H (2008) Simulation of resonant interaction between energetic electrons and whistler-mode chorus in the outer radiation belt. Chin Phys Lett 25:4493–4496

Su Z, Zheng H (2009) Resonant scattering of relativistic outer zone electrons by plasmaspheric plume electromagnetic ion cyclotron waves. Chin Phys Lett 26(129):401. doi:10.1088/0256-307X/26/12/129401

Su Z, Zheng H, Wang S (2009a) Dynamic evolution of energetic outer zone electrons due to whistler-mode chorus based on a realistic density model. J Geophys Res 114(A07):201. doi: 10.1029/2008JA014013

Su Z, Zheng H, Wang S (2009b) Evolution of electron pitch angle distribution due to interactions with whistler-mode chorus following substorm injections. J Geophys Res 114(A08):202. doi: 10.1029/2009JA014269
Su Z, Zheng H, Xiong M (2009c) Dynamic evolution of outer radiation belt electrons due to whistler-mode chorus. Chin Phys Lett 26(039):401
Su Z, Xiao F, Zheng H, Wang S (2010a) STEERB: a three-dimensional code for storm-time evolution of electron radiation belt. J Geophys Res 115(A09):208. doi:10.1029/2009JA015210
Su Z, Zheng H, Wang S (2010b) A parametric study on the diffuse auroral precipitation by resonant interaction with whistler-mode chorus. J Geophys Res 115(A05):219. doi:10.1029/2009JA014759
Su Z, Zheng H, Wang S (2010c) Three dimensional simulations of energetic outer zone electron dynamics due to wave-particle interaction and azimuthal advection. J Geophys Res 115(A06):203. doi:10.1029/2009JA014980
Su Z, Xiao F, Zheng H, Wang S (2011a) CRRES observation and STEERB simulation of the 9 October 1990 electron radiation belt dropout event. Geophys Res Lett 38(L06):106. doi:10.1029/2011GL046873
Su Z, Xiao F, Zheng H, Wang S (2011b) Radiation belt electron dynamics driven by adiabatic transport, radial diffusion, and wave-particle interactions. J Geophys Res 116(A04):205. doi: 10.1029/2010JA016228
Su Z, Zheng H, Chen L, Wang S (2011c) Numerical simulations of storm-time outer radiation belt dynamics by wave-particle interactions including cross diffusion. J Atmos Sol Terr Phys 73:95–105. doi:10.1016/j.jastp.2009.08.002
Subbotin D, Shprits Y, Ni B (2010) Three-dimensional VERB radiation belt simulations including mixed diffusion. J Geophys Res 115(A03):205. doi:10.1029/2009JA015070
Summers D (2005) Quasi-linear diffusion coefficients for field-aligned electromagnetic waves with applications to the magnetosphere. J Geophys Res 110(A08):213. doi:10.1029/2005JA011159
Summers D, Ma C (2000) A model for generating relativistic electrons in the Earth's inner magnetosphere based on gyroresonant wave-particle interactions. J Geophys Res 105:2625
Summers D, Omura Y (2007) Ultra-relativistic acceleration of electrons in planetary magnetospheres. Geophys Res Lett 34(L24):205. doi:10.1029/2007GL032226
Summers D, Thorne RM (2003) Relativistic electron pitch-angle scattering by electromagnetic ion cyclotron waves during geomagnetic storms. J Geophys Res 108:1143. doi:10.1029/2002JA009489
Summers D, Thorne RM, Xiao F (1998) Relativistic theory of wave-particle resonant diffusion with application to electron acceleration in the magnetosphere. J Geophys Res 103(20):487
Summers D, Ni B, Meredith NP (2007a) Timescales for radiation belt electron acceleration and loss due to resonant wave-particle interactions: 1. Theor J Geophys Res 112(A04):206. doi: 10.1029/2006JA011801
Summers D, Ni B, Meredith NP (2007b) Timescales for radiation belt electron acceleration and loss due to resonant wave-particle interactions: 2. Evaluation for VLF chorus, ELF hiss, and electromagnetic ion cyclotron waves. J Geophys Res 112(A04):207. doi:10.1029/2006JA011993
Summers D, Ni B, Meredith NP, Horne RB, Thorne RM, Moldwin MB, Anderson RR (2008) Electron scattering by whistler-mode elf hiss in plasmaspheric plumes. J Geophys Res 113(A04):219. doi:10.1029/2007JA012678
Tao X, Chan AA, Albert JM, Miller JA (2008) Stochastic modeling of multidimensional diffusion in the radiation belts. J Geophys Res 113(A07):212. doi:10.1029/2007JA012985
Tao X, Albert JM, Chan AA (2009) Numerical modeling of multidimensional diffusion in the radiation belts using layer methods. J Geophys Res 114(A02):215. doi:10.1029/2008JA013826
Tao X, Bortnik J, Albert JM, Liu K, Thorne RM (2011a) Comparison of quasilinear diffusion coefficients for parallel propagating whistler mode waves with test particle simulations. Geophys Res Lett 380(L06):105. doi:10.1029/2011GL046787
Tao X, Thorne RM, Li W, Ni B, Meredith NP, Horne RB (2011b) Evolution of electron pitch angle distributions following injection from the plasma sheet. J Geophys Res 116(A4): A04,229

Thorne RM (2010) Radiation belt dynamics: the importance of wave-particle interactions. Geophys Res Lett 37(L22):107. doi:10.1029/2010GL044990

Thorne RM, Smith EJ, Burton RK, Holzer RE (1973) Plasmaspheric hiss. J Geophys Res 78: 1581–1596. doi:10.1029/JA078i010p01581

Thorne RM, Ni B, Tao X, Horne RB, Meredith NP (2010) Scattering by chorus waves as the dominant cause of diffuse auroral precipitation. Nature 467:943. doi:10.1038/nature09467

Tsurutani BT, Smith EJ (1974) Postmidnight chorus: a substorm phenomenon. J Geophys Res 79:118–127. doi:10.1029/JA079i001p00118

Tsurutani BT, Smith EJ (1977) Two types of magnetospheric ELF chorus and their substorm dependences. J Geophys Res 82:5112–5128. doi:10.1029/JA082i032p05112

Varotsou A, Boscher D, Bourdarie S, Horne RB, Glauert SA, Meredith NP (2005) Simulation of the outer radiation belt electrons near geosynchronous orbit including both radial diffusion and resonant interaction with whistler-mode chorus waves. Geophys Res Lett 32(L19):106. doi: 10.1029/2005GL023282

Varotsou A, Boscher D, Bourdarie S, Horne RB, Meredith NP, Glauert SA, Friedel RH (2008) Three-dimensional test simulations of the outer radiation belt electron dynamics including electron-chorus resonant interactions. J Geophys Res 113(A12):212. doi:10.1029/2007JA012862

Vasyliunas VM (1968) A survey of low-energy electrons in the evening sector of the magnetosphere with OGO 1 and OGO 3. J Geophys Res 73:2839

Viñas AF, Mace RL, Benson RF (2005) Dispersion characteristics for plasma resonances of Maxwellian and Kappa distribution plasmas and their comparisons to the IMAGE/RPI observations. J Geophys Res 110(A06):202. doi:10.1029/2004JA010967

Xiao F (2006) Modelling energetic particles by a relativistic kappa loss cone distribution function in plasmas. Plasma Phys Controlled Fusion 48:203

Xiao F, He HY (2006) Energetic electron pitch angle diffusion due to whistler wave during terrestrial storms. Chin Phys Lett 23:267–270. doi:10.1088/0256-307X/23/1/077

Xiao F, Chen L, Zhou Q, He H, Wen Y (2007a) Bounce-averaged acceleration of energetic electrons by whistler mode chorus in the magnetosphere. Chin Phys Lett 24:294–297. doi:10.1088/0256-307X/24/1/080

Xiao F, Thorne RM, Summers D (2007b) Higher-order gyroresonant acceleration of electrons by superluminous (AKR) wave-modes. Planet Space Sci 55:1257

Xiao F, Chen L, Li J (2008a) Energetic particles modeled by a generalized relativistic kappa-type distribution function in plasmas. Plasma Phys Controlled Fusion 50(105):002. doi:10.1088/0741-3335/50/10/105002

Xiao F, Shen C, Wang Y, Zheng H, Wang S (2008b) Energetic electron distributions fitted with a relativistic kappa-type function at geosynchronous orbit. J Geophys Res 113(A05):203. doi: 10.1029/2007JA012903

Xiao F, Zhou Q, Li C, Cai A (2008c) Modeling solar energetic particle by a relativistic kappa-type distribution. Plasma Phys Controlled Fusion 50(062):001. doi:10.1088/0741-3335/50/6/062001

Xiao F, Su Z, Zheng H, Wang S (2009) Modeling of outer radiation belt electrons by multidimensional diffusion process. J Geophys Res 114(A03):201. doi:10.1029/2008JA013580

Xiao F, Chen L, Su Z, Zheng H, Wang S (2010a) A parametric study on outer radiation belt electron evolution by superluminous R-X mode waves. J Geophys Res 115(A10):217. doi: 10.1029/2010JA015374

Xiao F, Su Z, Zheng H, Wang S (2010b) Three-dimensional simulations of outer radiation belt electron dynamics including cross diffusion terms. J Geophys Res 115(A05):216. doi:10.1029/2009JA014541

Xiao F, Chen L, He Y, Su Z, Zheng H (2011) Modeling of precipitation loss of ring current protons by electromagnetic ion cyclotron waves. J Atmos Sol Terr Phys 73:106–111. doi:10.1016/j.jastp.2010.01.007

Yoon PH (2011) Large-amplitude whistler waves and electron acceleration. Geophys Res Lett 381(L12):105. doi:10.1029/2011GL047893

Zheng H, Su Z, Xiong M (2008) Pitch angle distribution evolution of energetic electrons by whistler-mode chorus. Chin Phys Lett 25:3515–3518. doi:10.1088/0256-307X/25/9/113
Zheng Q, Fok M, Albert J, Horne RB, Meredith NP (2011) Effects of energy and pitch angle mixed diffusion on radiation belt electrons. J Atmos Sol Terr Phys 73:785–795. doi:10.1016/j.jastp.2011.01.014

Chapter 3
Radial Diffusion

3.1 Introduction

The trapped particles in the inner magnetosphere have three adiabatic invariants associated to the three periodic motions (see Sect. 1.3.1). The ultra low-frequency (ULF) waves can drift-resonate with the energetic electrons, violate the third adiabatic invariant, and then drive the so-called radial diffusion process (Elkington et al. 1999, 2003; Hudson et al. 2000). The inward radial diffusion is an important acceleration mechanism for radiation belt electrons, while the outward radial diffusion can contribute to the loss of radiation belt electrons. The quiet-time steady distribution of radiation belt electrons derived from the radial diffusion theory was found to agree well with the observations (Lyons and Thorne 1973; Su et al. 2010a, b, 2011a, b). Li et al. (2001) showed that the radial diffusion theory with diffusion coefficients depending on interplanetary parameters can well predict the energetic electron flux evolution in the geosynchronous orbit. However, there have been increasing storm-time case studies (e.g., Brautigam and Albert 2000; Horne et al. 2005; Miyoshi et al. 2006; Shprits et al. 2006; Albert et al. 2009) suggesting that the radial diffusion is insufficient to reproduce the rate and extent of energetic electron flux enhancements during the storm recovery phase in the absence of an additional energization source.

Radiation belt models are required to understand the physical mechanisms for its sources and sinks, simulate its configuration, and predict its variability. Beutier and Boscher (1995) developed the Salammbô code via solving the three-dimensional Fokker-Planck equation, simultaneously treating the deceleration of radiation belt electrons by free and bounded thermospheric and ionospheric electrons, pitch-angle diffusion by Coulomb and plasmaspheric hiss wave-particle interactions, radial diffusion by magnetic and electric field perturbations, and cosmic ray albedo neutron decay process. Bourdarie et al. (1997) presented the first four-dimensional model for storm-time electron radiation belt dynamics with incorporation of the deceleration by free and bounded electrons of the medium, pitch-angle diffusion by Coulomb and wave-particle interactions, radial and azimuthal transport. Varotsou et al. (2005, 2008) introduced the low-latitude chorus wave-particle interactions out

Z. Su, *A Global Kinetic Model for Electron Radiation Belt Formation and Evolution*, Springer Theses, DOI 10.1007/978-3-662-46651-3_3

of plasmapause into the Salammbô code. Jordanova and Miyoshi (2005), Miyoshi et al. (2006), Jordanova et al. (2008) proposed the four-dimensional RAM model considering the convective transport of electrons, radial diffusion, and wave-particle interactions. Fok et al. (2008) developed the four-dimensional RBE model including the magnetospheric convection, wave-particle interaction processes. Shprits et al. (2008), Subbotin and Shprits (2009) constructed the three-dimensional VERB code based on the previous one-dimensional radial diffusion code (Shprits and Thorne 2004) and two-dimensional wave-particle interaction code (Shprits et al. 2006). In a follow-on study, Shprits et al. (2009) presented some idealized simulations of storm-time electron radiation belt dynamics using VERB code including the pitch-angle and energy diffusion due to chorus, plasmaspheric hiss, plume hiss and EMIC waves, and radial diffusion due to magnetic field perturbations. Albert et al. (2009) simulated the outer radiation belt electrons dynamics during the 9 October 1990 magnetic storm using the three-dimensional modified Fokker-Planck equation including radial diffusion due to magnetic and electric field perturbations, and pitch-angle, energy and cross-pitch-angle-energy diffusion due to chorus wave-particle interactions.

In this chapter, we present a detailed description of the three-dimensional code for Storm-Time Evolution of Electron Radiation Belt (STEERB). The STEERB code fully takes into account the radial diffusion and the local diffusion driven by cyclotron resonance. Based on the STEERB code, we further analyze the steady distribution of radiation belt electrons, and quantify the respective contributions of radial diffusion and local diffusion to the radiation belt evolution. Particularly, considering that the cross-pitch-angle-momentum diffusion is often ignored in the previous models, we here evaluate the importance of cross-term in the simulations.

3.2 Global Radiation Belt Diffusion Model STEERB

3.2.1 Background Magnetic Field

The geomagnetic field here is assumed to be the typical dipole magnetic field. The dynamic geomagnetic field will be introduced in the next chapter.

3.2.2 Basic Equation

The three-dimensional diffusion equation for the radiation belt electron PSD f evolution can be written as (e.g., Schulz and Lanzerotti 1974; Beutier and Boscher 1995; Varotsou et al. 2005, 2008; Subbotin and Shprits 2009; Su et al. 2010a; Xiao et al. 2010)

$$
\begin{aligned}
\frac{\partial f}{\partial t} =& \frac{1}{Gp} \frac{\partial}{\partial \alpha_e}\bigg|_{p,L} \left[G \left(\langle D_{\alpha\alpha} \rangle \frac{1}{p} \frac{\partial f}{\partial \alpha_e}\bigg|_{p,L} + \langle D_{\alpha p} \rangle \frac{\partial f}{\partial p}\bigg|_{\alpha_e,L} \right) \right] \\
&+ \frac{1}{G} \frac{\partial}{\partial p}\bigg|_{\alpha_e,L} \left[G \left(\langle D_{p\alpha} \rangle \frac{1}{p} \frac{\partial f}{\partial \alpha_e}\bigg|_{p,L} + \langle D_{pp} \rangle \frac{\partial f}{\partial p}\bigg|_{\alpha_e,L} \right) \right] \\
&+ L^2 \frac{\partial}{\partial L}\bigg|_{\mu,J} \left(\frac{D_{LL}}{L^2} \frac{\partial f}{\partial L}\bigg|_{\mu,J} \right) - \frac{f}{\tau_L} - \frac{f}{\tau_C} - \frac{f}{\tau_W}.
\end{aligned}
\tag{3.1}
$$

On the right-hand side of this equation, the first two terms correspond to the local diffusion associated with the cyclotron resonance in the observable space (α_e, p, L) with the drift-averaged diffusion coefficients $\langle D_{\alpha\alpha} \rangle$, $\langle D_{pp} \rangle$ and $\langle D_{\alpha p} \rangle = \langle D_{p\alpha} \rangle$ (see Chap. 2). The third term corresponds to the radial diffusion in the adiabatic invariant space (μ, J, L) with the diffusion coefficient D_{LL}; The last three terms represent the losses in the loss cone, near the Earth ($L < 1.5$) and in the slot region: τ_L is the electron lifetime related to precipitation loss, which is set to be a quarter of bouncing time in the loss cone $\alpha_L = \arcsin\left[L^5(4L-3)\right]^{-1/4}$ and infinite outside the loss cone (Shprits et al. 2009); τ_C is the electron lifetime related to Coulomb scattering (Wentworth et al. 1959; Lyons and Thorne 1973), which is set to be

$$
\tau_C = 3 \times 10^{11} E_k / N(L) \quad [\mathrm{s}],
\tag{3.2}
$$

with electron kinetic energy E_k in units of MeV, and particle number density $N(L) = 1000(4/L)^4$ in units of cm^{-3} (Komatsu and Watanabe 2008); τ_W is the electron lifetime related to the resonant scattering by plasmaspheric hiss, lightning-generated whistlers and VLF transmitter signals in the slot region, given by an empirical relation based on the previous works (Abel and Thorne 1998; Meredith et al. 2007)

$$
\tau_W = \left(\frac{2}{K_p} \right)^2 10^{(L-3.5)^4 + 0.4} \quad (2.5 \le L \le L_{pp}) \quad [\mathrm{day}],
\tag{3.3}
$$

where the plasmapause location L_{pp} is computed as (Carpenter and Anderson 1992)

$$
L_{pp} = 5.6 - 0.46 K_p^*,
\tag{3.4}
$$

with K_p^* is the maximum K_p value in the preceding 24 h. It should be noted that only the dependence of τ_W on L is considered, while the dependence on energy and pitch-angle is neglected. Electron lifetime τ_W in the energy range from ~0.1 to ~5 MeV is comparable with the previous studies (Abel and Thorne 1998; Meredith et al. 2007).

In such a dipole magnetic field, the mapping relation between the two computational space is expressed as

$$\mu = \frac{p^2 \sin^2 \alpha_e L^3}{2m_e B_0}, \tag{3.5a}$$

$$J = 2pLR_E Y(\sin \alpha_e), \tag{3.5b}$$

$$L = L, \tag{3.5c}$$

More explanations can be found in the Sect. 1.3.1.

3.2.3 Diffusion Coefficients

We have developed the code for the local diffusion coefficients in the Chap. 2. In the present chapter, we take into account the contributions of chorus, hiss, and EMIC waves. The chorus waves are distributed in the plasmatrough, with the electron density (Sheeley et al. 2001)

$$N_e(L) = 124 \left(\frac{3}{L}\right)^4 \quad [\mathrm{cm}^{-3}]. \tag{3.6}$$

The hiss and EMIC waves are distributed in the plasmaspheric plume, with the electron density (Carpenter and Anderson 1992)

$$N_e(L) = 10^{-0.3145L + 3.9043} \quad [\mathrm{cm}^{-3}]. \tag{3.7}$$

All the required wave parameters are listed in the Table 3.1 (Shprits et al. 2009). Figure 3.1 shows the two-dimensional distribution of bounce-averaged diffusion coefficients for chorus, plume hiss, and EMIC waves at $L = 4.5$, generally consistent with the previous studies (Xiao et al. 2009; Su et al. 2011c). Some two-dimensional local diffusion simulations have been presented to study the role of each waves (e.g., Li et al. 2007; Su et al. 2011c). The dayside and nightside chorus waves can produce the local acceleration of energetic electrons especially at large pitch-angles. The hiss wave is able to efficiently redistribute electrons along pitch-angle and drive the near-equatorially mirroring electrons toward the loss cone. EMIC waves can rapidly scatter the energetic electrons into loss cone and produce precipitation loss to atmosphere.

The radial diffusion coefficient D_{LL} consists of magnetic field perturbation D_{LL}^M and electric field perturbation D_{LL}^E, modeled by (Brautigam and Albert 2000; Albert et al. 2009),

$$D_{LL}^M = 10^{0.506K_p - 9.325} L^{10} \quad [1/\mathrm{day}], \tag{3.8}$$

$$D_{LL}^E = \frac{c^2 \widetilde{E}^2}{4R_e^2 B_g^2} \frac{T_d}{1 + (\omega_d T_d/2)^2} L^6 \quad [1/\mathrm{s}] \quad \text{(in Gaussian units)}, \tag{3.9}$$

Table 3.1 Parameters for the storm-time properties of chorus, hiss, and EMIC waves

Waves	Dayside chorus	Nightside chorus	Hiss	EMIC
B_t (pT)	$10^{0.75+0.04\lambda} \times \left(\frac{2}{3391.2} 10^{2.5+0.18K_p}\right)^{0.5}$	$50 \times \left(\frac{2}{3391.2} 10^{2.5+0.18K_p}\right)^{0.5}$	$30 \times K_p/4$	1000
λ_m (Deg)	35	15	40	15
P (%)	25	25	15	5
Spectral property	$\omega_1/\|\Omega_e\|=0.10$ $\omega_2/\|\Omega_e\|=0.30$ $\omega_m/\|\Omega_e\|=0.20$ $\delta\omega/\|\Omega_e\|=0.10$	$\omega_1/\|\Omega_e\|=0.05$ $\omega_2/\|\Omega_e\|=0.65$ $\omega_m/\|\Omega_e\|=0.35$ $\delta\omega/\|\Omega_e\|=0.15$	$\omega_1 = 0.0628 \times 10^4$ rad/s $\omega_2 = 1.2500 \times 10^4$ rad/s $\omega_m = 0.3580 \times 10^4$ rad/s $\delta\omega = 0.1790 \times 10^4$ rad/s	$\omega_1/\|\Omega_{O+}\|=3.45$ $\omega_2/\|\Omega_{O+}\|=3.95$ $\omega_m/\|\Omega_{O+}\|=3.70$ $\delta\omega/\|\Omega_{O+}\|=0.25$
Normal angle distribution	$X_1=0$ $X_2=1$ $X_m=0$ $X_\omega=0.577$	$X_1=0$ $X_2=1$ $X_m=0$ $X_\omega=0.577$	$X_1=0$ $X_2=1$ $X_m=0$ $X_\omega=0.577$	Parallel propagation approximation

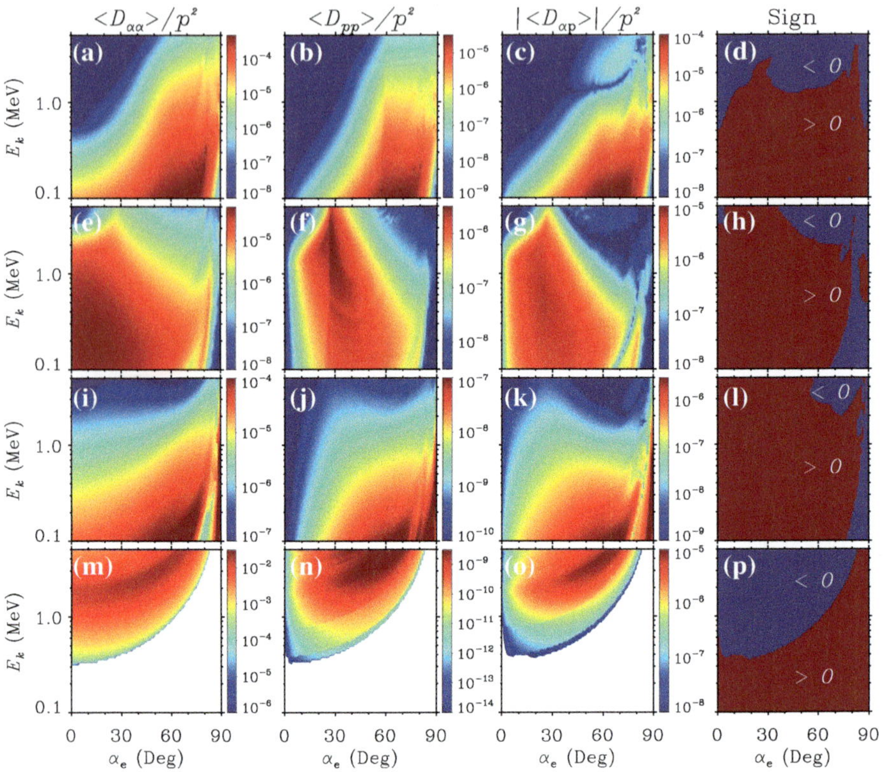

Fig. 3.1 Two-dimensional distribution of bounce-averaged diffusion coefficients as functions of pitch-angle α_e and kinetic energy E_k for nightside chorus (*first row*), dayside chorus (*second row*), hiss (*third row*), EMIC waves (*fourth row*) at $L = 4.5$ in units of s^{-1}

where the electric field fluctuation $\widetilde{E} = \widetilde{E}_0 + \widetilde{E}_1(K_p - 1)$ with $\widetilde{E}_0 = 3.33 \times 10^{-9}$ stat V/cm (0.1 mV/m) and $\widetilde{E}_1 = 8.67 \times 10^{-9}$ stat V/cm (0.26 mV/m), fluctuation decay time $T_d = 2700$ s, and the drift frequency (Schulz et al. 1991)

$$\omega_d = \frac{3m_e c^2 \gamma \beta^2 L}{eB_g R_e^2} \frac{D(\alpha_e)}{T(\alpha_e)}$$
$$\approx 1.2 \times 10^{-3} \gamma \beta^2 L \frac{D(\alpha_e)}{T(\alpha_e)} \text{ [rad/s]}, \tag{3.10}$$

with the speed of light c, electron charge e, relativistic Lorentz factor γ, $\beta = \sqrt{1 - \frac{1}{\gamma^2}}$, and $D(\alpha_e) = (5.520692 - 2.357194 \sin\alpha_e + 1.279385 \sin\alpha_e^{3/4})/12$ (Schulz et al. 1991). Note here that the expression of drift frequency (3.10) can degenerate into that adopted by Brautigam and Albert (2000), Albert et al. (2009) when $\alpha_e = 90°$. Figure 3.2 shows the profiles of radial diffusion coefficients D_{LL}^M and D_{LL}^E, electron lifetimes τ_C and τ_W during quite time ($K_p = 2$). These timescales

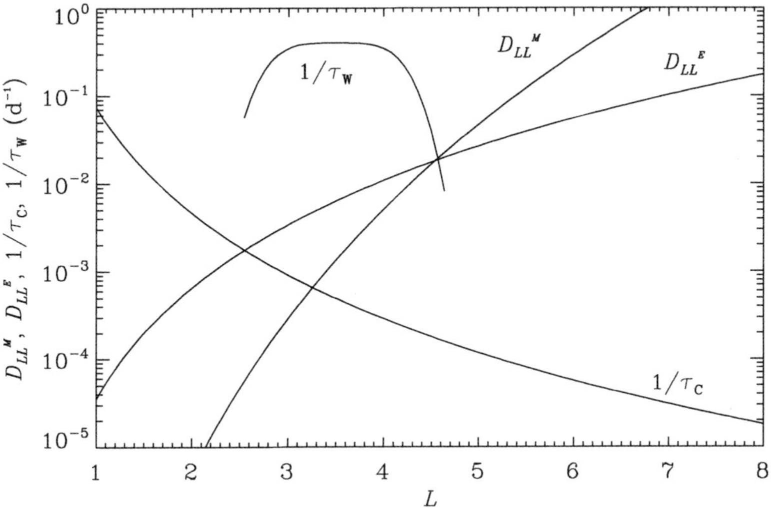

Fig. 3.2 Profiles of radial diffusion coefficients and electron lifetimes for equatorially-trapped electrons with energies $E_k = 1.0\,\text{MeV}$ during quite time ($K_p = 2$)

are found to be generally comparable with those in the earlier study (Beutier and Boscher 1995).

These expressions for radial diffusion coefficients D_{LL}^M and D_{LL}^E are obtained from the observation data in outer radiation belt, and it is questionable to apply them to the slot and inner belt regions (Komatsu and Watanabe 2008). In fact, the accurate dependence of radial diffusion coefficients on L still needs to be further determined, considering the uncertainty of actual dependence of ULF wave power on L (Fei et al. 2006; Sarris et al. 2006; Li et al. 2009). Until better information exists, the overall radial diffusion coefficient is therefore adjusted by two free parameters, P_m and P_e:

$$D_{LL} = P_m D_{LL}^M + P_e D_{LL}^E. \tag{3.11}$$

3.2.4 Numerical Method

Based on the split operator technique (Strang 1968), numerical solutions of the three-dimensional diffusion Eq. (3.1) are carried out in two steps associated with the local diffusion

$$\frac{\partial f}{\partial t} = \frac{1}{G}\frac{\partial}{\partial \alpha_e}\bigg|_{p,L}\left[G\left(\langle D_{\alpha\alpha}\rangle \frac{\partial f}{\partial \alpha_e}\bigg|_{p,L} + \langle D_{\alpha p}\rangle \frac{\partial f}{\partial p}\bigg|_{\alpha_e,L}\right)\right] + \frac{1}{G}\frac{\partial}{\partial p}\bigg|_{\alpha_e,L}\left[G\left(\langle D_{p\alpha}\rangle \frac{\partial f}{\partial \alpha_e}\bigg|_{p,L} + \langle D_{pp}\rangle \frac{\partial f}{\partial p}\bigg|_{\alpha_e,L}\right)\right] - \frac{f}{\tau_L} \tag{3.12}$$

and the radial diffusion

$$\frac{\partial f}{\partial t} = L^2 \left.\frac{\partial}{\partial L}\right|_{\mu,J} \left(\frac{D_{LL}}{L^2} \left.\frac{\partial f}{\partial L}\right|_{\mu,J} \right) - \frac{f}{\tau_C} - \frac{f}{\tau_W}. \tag{3.13}$$

Each of these equations are solved once, then again in reverse order for next time step to diminish systematic error.

The local diffusion process (3.12) is implemented using the HFD method (Xiao et al. 2009) (more details in the Chap. 2) in the (α_e, p, L) space, which is found to be stable and efficient even on a large computational domain for a long time (Su et al. 2009a, b, c, 2010c, d, 2011c). The computational domain covers $[0°, 90°] \times [0.1\,\text{MeV}, 5\,\text{MeV}] \times [1, 8]$ in the (α_e, E_k, L) space. The grid points are orthogonal in the (α_e, p, L) space. We use uniform grids in the pitch-angle α_e and Roederer parameter L, and natural logarithmic grids in the momentum p. The corresponding grid numbers set to be $91 \times 81 \times 71$ in the (α_e, p, L) space, possessing high spatial resolutions. The following boundary conditions are adopted to implement the local diffusion operators

$$\partial f/\partial \alpha_e(\alpha_e = 0°) = 0 \qquad \partial f/\partial \alpha_e(\alpha_e = 90°) = 0 \tag{3.14}$$

$$f(E_k = 0.1\,\text{MeV}) = f_l(t, \alpha_e, L) \quad f(E_k = 5\,\text{MeV}) = f_u(t, \alpha_e, L), \tag{3.15}$$

where $f_l(t, \alpha_e, L)$ and $f_u(t, \alpha_e, L)$ are given self-consistently by the interpolations from (μ, J, L) space at each step.

The radial diffusion Eq. (3.13) is implemented via an unconditionally stable, fully implicit finite difference method in the (μ, J, L) space. The grids are evenly distributed in the direction L with grid number 71, while nonuniform grids are adopted in the $\mu \times J$ plane at a constant L. At $L = 8$, the mapped grids are orthogonal in the (α_e, p) plane, and cover $[0°, 90°] \times [0.1\,\text{keV}, 5\,\text{MeV}]$ in the (α_e, E_k) plane. We adopt the uniform grids in pitch-angle α_e, and natural logarithmic grids in momentum p at $L = 8$. The corresponding grid numbers are 91×201 in the (α_e, p) plane. The α_e and p values of mapped grids for other L shells can be obtained from the conservation of adiabatic invariants. The following boundary conditions are adopted to implement the radial diffusion operators

$$f(L = 1) = 0, \ f(L = 8) = f_o(t, \alpha_e, p), \tag{3.16}$$

considering that losses dominate at $L = 1$, and the time-dependent source $f_o(t, \alpha_e, p)$ at $L = 8$ can be given by observation or other empirical distribution.

Interpolations between the two spaces are required since the mapped grids are not regularly aligned. The finite difference methods for both local wave-particle interaction (3.12) and radial diffusion (3.13) process generally possess second-order accuracy in space, and then cubic spline interpolation methods are adopted in order to keep the second-order space accuracy of total scheme. In addition, Subbotin and Shprits (2009) showed that spline interpolation can effectively keep the shape of interpolated function without introducing visible errors even for a long-time computation.

The numerical implementation involves a series of intrinsically concurrent interpolations and solutions of tridiagonal systems of linear equations at each step. Hence, parallelization of this numerical scheme is easy to be realized via the OpenMP or MPI interface.

3.3 Idealized Simulations

3.3.1 Steady State

The stable configuration of radiation belt can be obtained by solving the steady radial diffusion Eq. (4.17) with outer boundary condition given by a kappa-type distribution function (Vasyliunas 1968; Maksimovic et al. 1997a, b; Viñas et al. 2005; Xiao 2006; Xiao et al. 2008a, b, c):

$$f^{\kappa}(\alpha_e, E_k) = c_{\kappa}\left[1 + \frac{E_k}{\kappa E_0}\right]^{-(\kappa+1)} \sin^q \alpha_e \tag{3.17}$$

where c_{κ} is a normalized constant, $q = 0.2$, $E_0 = 2\,\text{keV}$, and $\kappa = 5$, are the characteristics of the plasmasheet (Christon et al. 1988; Thorne et al. 2005). Lots of test simulations are executed, and the optimal parameters for radial diffusion coefficients would be $P_m = 1.0$ and $P_e = 0.2$. The magnetic perturbation D_{LL}^{M} tends to dominant throughout the outer radiation belt, while the electric perturbations are dominant in the slot region and inner belt. Stronger electric perturbation can lead to the filling of slot region and extremely large electron flux near the Earth.

The obtained steady-state flux profiles during quite time ($K_p = 2.0$) are shown in Fig. 3.3. The flux profiles observed by MEA instrument at Orbit 182 of CRRES satellite with energy channels 0.509 and 1.090 MeV (Thorne et al. 2007) are also plotted for comparison. The obtained stable configuration, inner and outer belts separated by slot region ($2{-}3R_e$), can be clearly distinguished, generally comparable with the previous observations (Thorne et al. 2007).

3.3.2 Storm-Time Dynamics

Following the previous work (Shprits and Thorne 2004), the time-dependent K_p index of an idealized storm is assumed to be

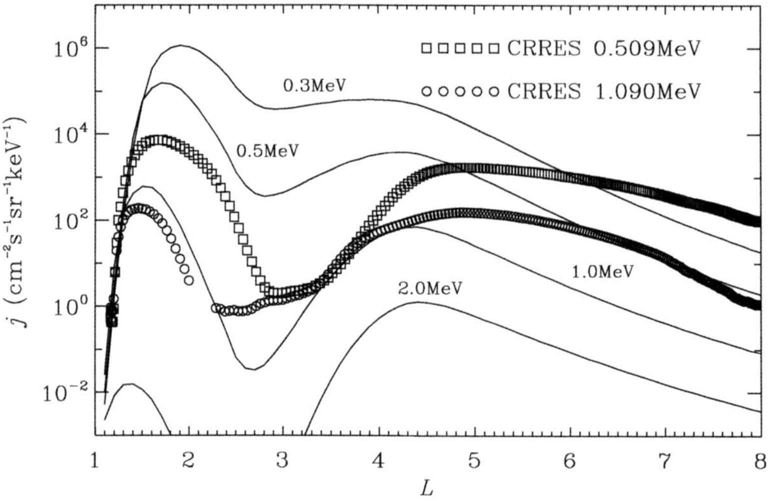

Fig. 3.3 Steady-state flux profiles (*solid lines*) of equatorially-trapped electrons with energies $E_k = 0.3, 0.5, 1.0$, and 2.0 MeV during quite time ($K_p = 2.0$). The *squares* and *circles* represent flux profiles observed by MEA instrument at Orbit 182 of CRRES satellite with energy channels 0.509 and 1.090 MeV (Thorne et al. 2007)

$$K_p = \begin{cases} 4t + 2, & 0.0 \le t < 1.0 \text{ day} \\ (-4t + 22)/3, & 1.0 \le t < 4.0 \text{ days} \\ 2, & t \ge 4.0 \text{ days} \end{cases}. \tag{3.18}$$

The outer boundary condition is given by

$$f_o(t, \alpha_e, p) = \overline{f}(t) f^{\kappa}(\alpha_e, E_k), \tag{3.19}$$

with

$$\overline{f}(t) = \begin{cases} 1, & 0.0 \le t < 0.5 \text{ day} \\ (-18t + 19)/10, & 0.5 \le t < 1.0 \text{ day} \\ (18t - 17)/10, & 1.0 \le t < 1.5 \text{ days} \\ 1, & t \ge 1.5 \text{ days} \end{cases}. \tag{3.20}$$

In order to investigate the effect of several important mechanisms, including radial diffusion (RD), cyclotron resonant diffusion by chorus waves (CW), as well as plume hiss and EMIC waves (PW), which may affect the global radiation belt dynamics, we perform three test simulation cases using STEERB code: (A) RD; (B) RD+CW; (C) RD+CW+PW. The time-dependent outer boundary condition

$$f_o(t, \mu, J) = \overline{f}(t) f_o(t = 0, \mu, J), \tag{3.21}$$

$$\overline{f}(t) = \begin{cases} 1, & 0.0 \leq t < 0.5\ \text{d} \\ (-18t + 19)/10, & 0.5 \leq t < 1.0\ \text{d} \\ (18t - 17)/10, & 1.0 \leq t < 1.5\ \text{d} \\ 1, & t \geq 1.5\ \text{d} \end{cases}. \tag{3.22}$$

is adopted in these cases. In the time range 0.5 d $< t <$ 1.5 d, $\overline{f}(t) < 1$ represents the loss of radiation belt electrons near the outer boundary (Shprits and Thorne 2004; Shprits et al. 2006). In order to further examine the effect of boundary loss, we further perform the fourth test simulation case (D) RD+CW+PW+CB with the constant boundary (CB) conditions. In Cases C and D, the plume waves are introduced over the spatial region $L_{pp} \leq L \leq 6.0$ only during the most geomagnetically active times (0.5–1.5 days).

Figure 3.4 shows the evolutions of equatorially-trapped electron PSD at $\mu = 450$ and 900 MeV/G for the four cases. The corresponding electron PSD profiles during

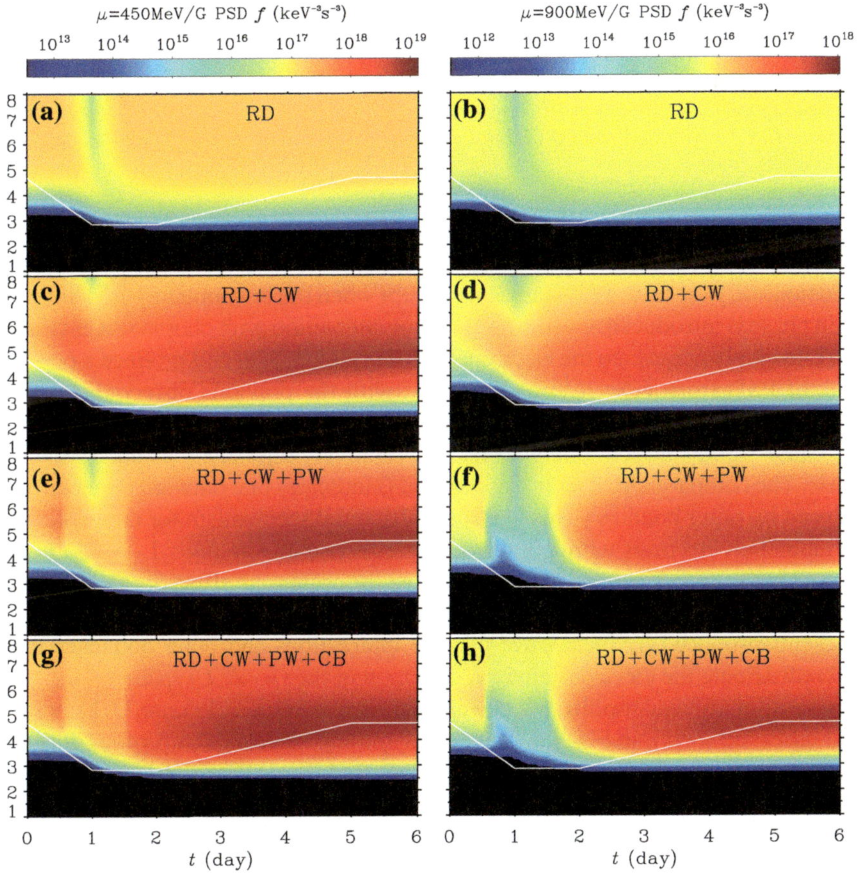

Fig. 3.4 Evolution of equatorially-trapped electron PSD at $\mu = 450$ (*left column*) and 900 MeV/G (*right column*) with **a** and **b** for Case A, **c** and **d** for Case B, **e** and **f** for Case C, **g** and **h** for Case D. The *white lines* denote the location of plasmapause L_{pp}

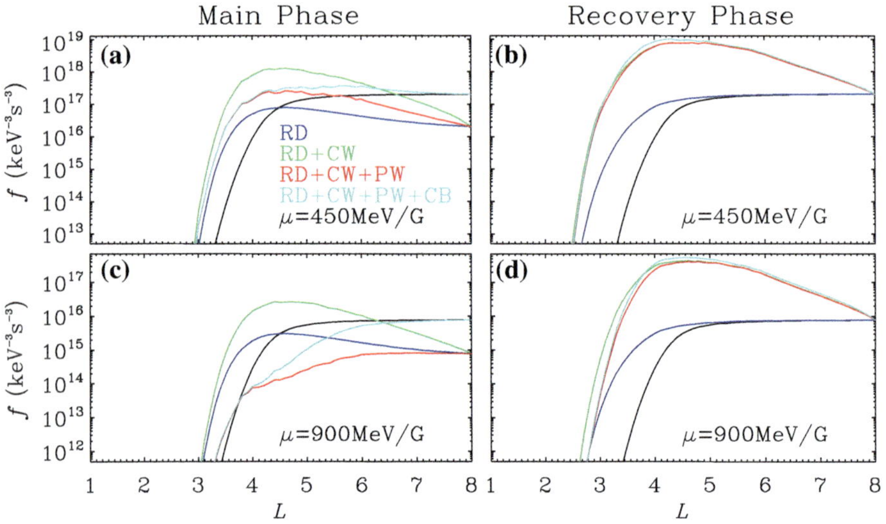

Fig. 3.5 Equatorially-trapped electron PSD profiles at $\mu = 450$ (*first row*) and 900 MeV/G (*second row*) during the main phase ($t = 1.0$ day) and recovery phase ($t = 4.0$ days) for Cases A (*dotted*), B (*dashed*), C (*dot-dashed*), and D (*dot-dot-dashed*). The *solid lines* represent the steady state

the main phase ($t = 1.0$ day) and recovery phase ($t = 4.0$ days) are shown in Fig. 3.5. If only the radial diffusion (Case A) is considered, electron PSD is found to monotonically increase as L increasing except when the out boundary loss during the main phase is introduced. Cyclotron resonant interaction with chorus waves (Case B) can effectively accelerate the electrons out of plasmapause, and produce a persistent peak of PSD from $L = 4$ to 6, consistent with the previous observations (Green and Kivelson 2004). Cyclotron resonant interactions with plume waves (Case C) can cause significant loss of electrons with relatively high energies during the main phase. The comparison between Cases C and D indicates that outward radial diffusion induced by outer boundary loss is also a substantial mechanism for the depletion of radiation belt electrons with arbitrary energies especially at larger L-shells ($L > 6$). These locally introduced acceleration and loss processes can affect the global electron radiation belt configuration, since the radial diffusion can smooth the electron PSD gradients produced by local wave-particle interactions and outer boundary losses, and redistribute the electrons along L direction in the (μ, J, L) space.

Evolutions of equatorially-trapped electron fluxes at $E_k = 1.0$ and 2.0 MeV for the four cases are shown in Fig. 3.6. The corresponding electron flux profiles during the main phase ($t = 1.0$ day) and recovery phase ($t = 4.0$ days) are plotted in Fig. 3.7. It is clear that the inner edge of outer radiation belt can penetrate into the slot region during the storms. Inward radial diffusion (Case A) can enhance the energetic electron flux in the range $2 < L < 5$ during active periods. Incorporation of cyclotron resonant interaction with chorus waves (Case B) causes more significant enhancement of energetic (~MeV) electron PSD over a wider L range especially during the recovery

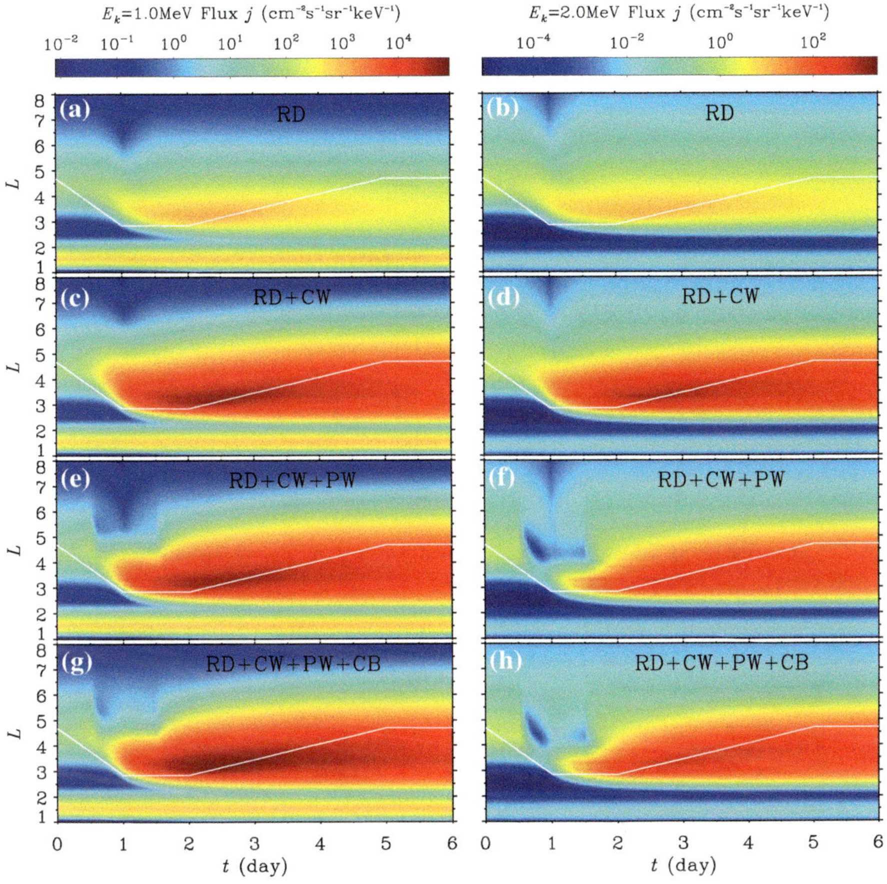

Fig. 3.6 The same as Fig. 3.4 except for electron fluxes at energies $E_k = 1.0$ (*left column*) and 2.0 MeV (*right column*)

phase. The cyclotron resonant interaction with plume hiss and EMIC waves (Case C) can yield notable energetic electrons losses in the range $L < 6$. The comparison between Cases C and D suggests that outward radial diffusion induced by outer boundary loss can lead to the energetic electron losses at larger L-shells ($L > 6$). In particular, when radial diffusion, cyclotron resonant interaction with chorus and plume waves as well as outer boundary losses are included (Case C), the energetic electron fluxes are found to decrease about one order during the main phase, and substantially increase by about three orders or above during the recovery phase in the heart of outer radiation belt, comparable with the previous observations (Reeves et al. 2003; Green et al. 2004).

Several two-dimensional (Albert and Young 2005; Tao et al. 2008, 2009; Xiao et al. 2009; Su et al. 2011c) and three-dimensional (Albert et al. 2009; Xiao et al. 2010) studies have been presented to investigate the effect of cross-pitch-angle-energy

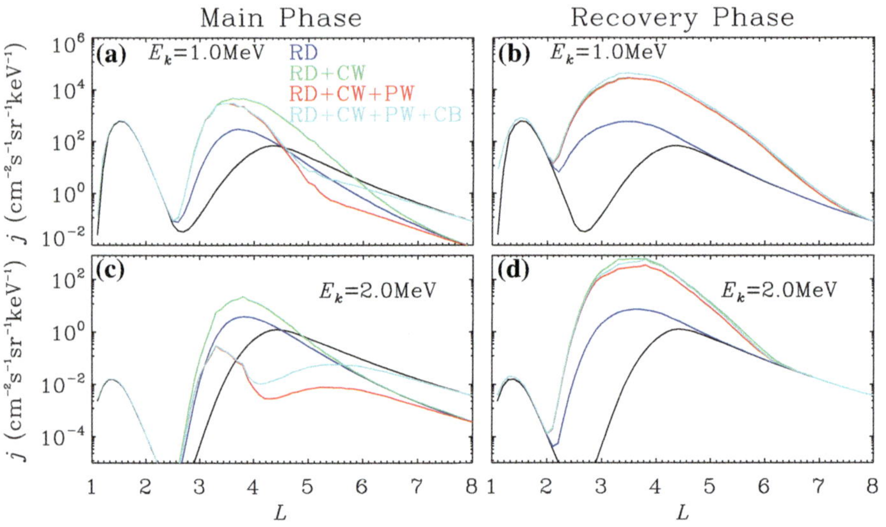

Fig. 3.7 The same as Fig. 3.5 except for electron fluxes at energies $E_k = 1.0$ (*first row*) and 2.0 MeV (*second row*)

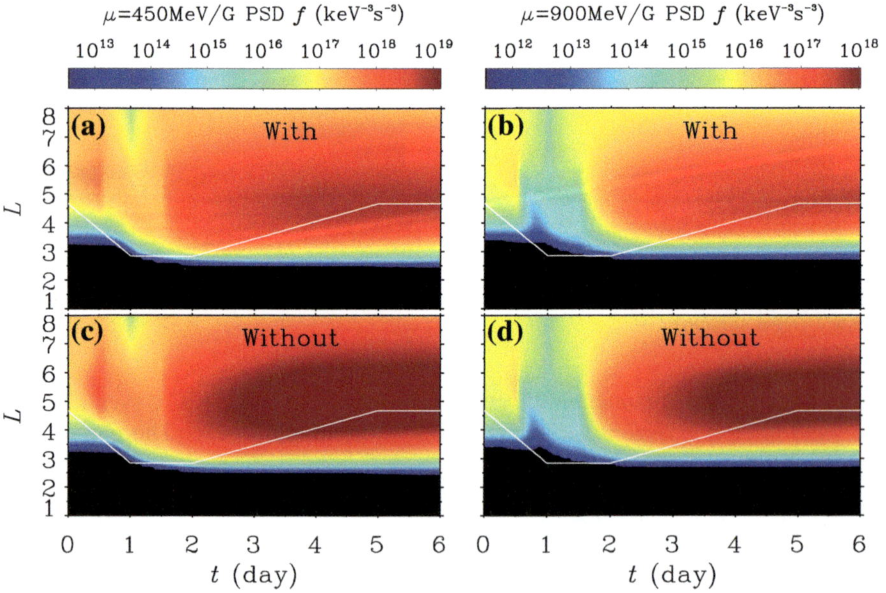

Fig. 3.8 Evolution of equatorially-trapped electron PSD at $\mu = 450$ (*left column*) and 900 MeV/G (*right column*) with (*first row*) and without (*second row*) cross-diffusion. The *white lines* denote the location of plasmapause L_{pp}

diffusion. Here, we will revisit the importance of cross-diffusion in three-dimensional simulations by comparison. Figure 3.8 shows the evolutions of equatorially-trapped

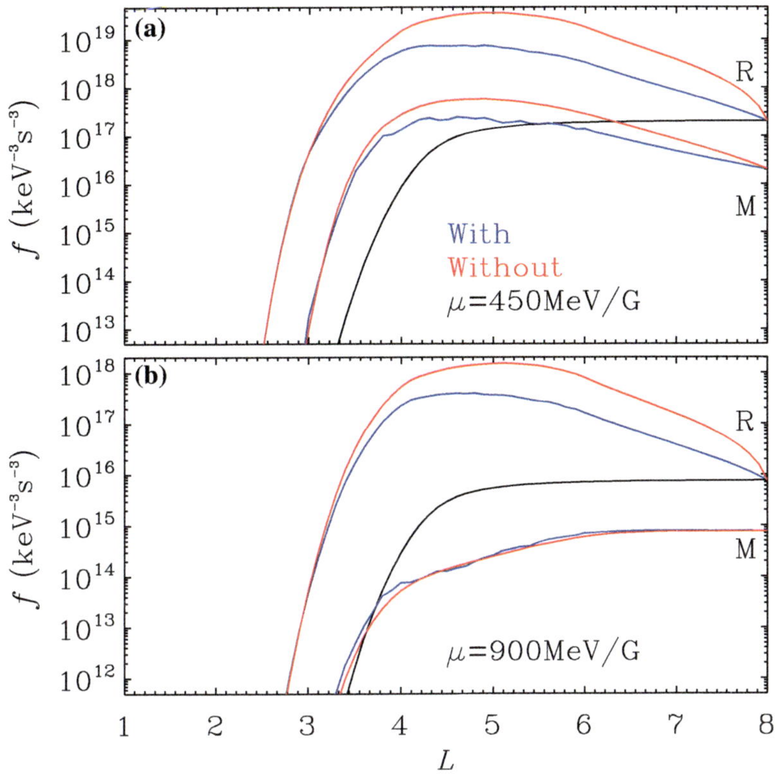

Fig. 3.9 Equatorially-trapped electron PSD profiles at μ = **a** 450 and **b** 900 MeV/G during the main phase (M, t = 1.0 day) and recovery phase (R, t = 4.0 days) with (*dotted*) and without (*dashed*) cross-diffusion. The *solid lines* represent the steady state

electron fluxes at μ = 450 and 900 MeV/G with and without incorporation of cross-diffusion. The corresponding electron flux profiles during the main phase (t = 1.0 day) and recovery phase (t = 4.0 days) are presented in Fig. 3.9. The equatorially-trapped electron PSD for μ =450 MeV/G during main phase is found be overestimated by a factor of 3 in the heart of outer radiation belt, and the overestimate factor can reach 10 during the recovery phase. The equatorially-trapped electron PSD for μ = 900 MeV/G during the main phase is almost unchanged when ignoring the cross-diffusion, while during the recovery phase it is found to be overestimated by a factor of 5−10 in the heart of outer radiation belt. These overestimate factors are comparable with previous three-dimensional studies (Xiao et al. 2010). It is obvious that these L-dependent PSD profiles with and without incorporation of cross-diffusion are similar to each other during both main and recovery phase, suggesting that neglect of cross-diffusion may not distort the dominant characteristics of PSD especially at large pitch-angles.

Figure 3.10 shows the evolutions of equatorially-trapped electron fluxes at $E_k = 1.0$ and 2.0 MeV with and without incorporation of cross-diffusion. The corresponding

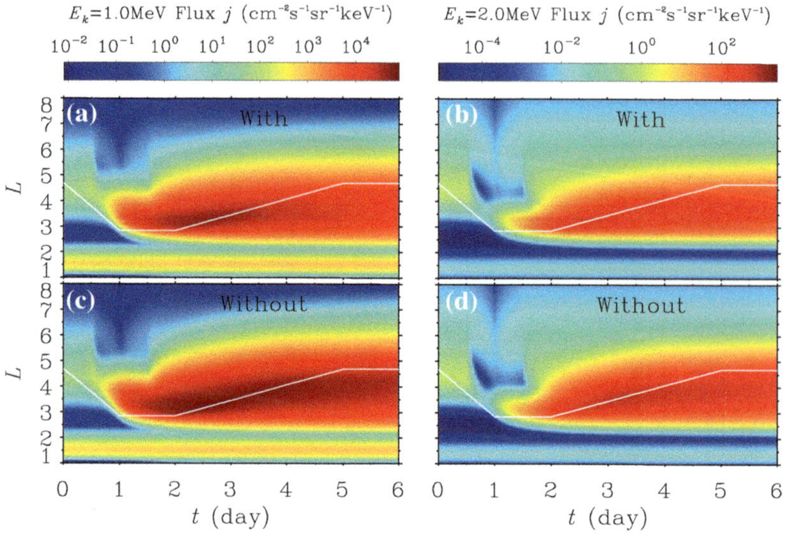

Fig. 3.10 The same as Fig. 3.8 except for electron fluxes at energies $E_k = 1.0$ (*left column*) and 2.0 MeV (*right column*)

electron flux profiles during the main phase ($t = 1.0$ day) and recovery phase ($t = 4.0$ days) are plotted in Fig. 3.11. During main phase, the difference caused by ignoring cross-diffusion is relatively small. During recovery phase, the equatorially-trapped energetic electron PSD can be overestimated by a factor of ∼5 in the heart of outer radiation belt, generally consistent with the previous two-dimensional local diffusion simulations (Albert and Young 2005; Tao et al. 2008, 2009; Su et al. 2011c).

3.4 Conclusions and Discussions

We present a three-dimensional code (STEERB) for the storm-time evolution of electron radiation belt via solving the three-dimensional diffusion equation. STEERB code covers the inner and outer radiation belt regions taking into account the Coulomb collisions, radial diffusion due to magnetic and electric field perturbations, and local pitch-angle, energy and cross-pitch-angle-energy diffusion due to various wave-particle interactions. STEERB is one of the first numerical models for storm-time electron radiation belt dynamics via solving the fully three-dimensional diffusion equation particularly including the cross-pitch-angle-anergy diffusion. It is implemented by a split operator technique, in conjunction with the recently developed HFD method for local wave-particle interaction, and the fully implicit finite difference method for radial diffusion. The resulting numerical model is robust, efficient, and easily parallelizable.

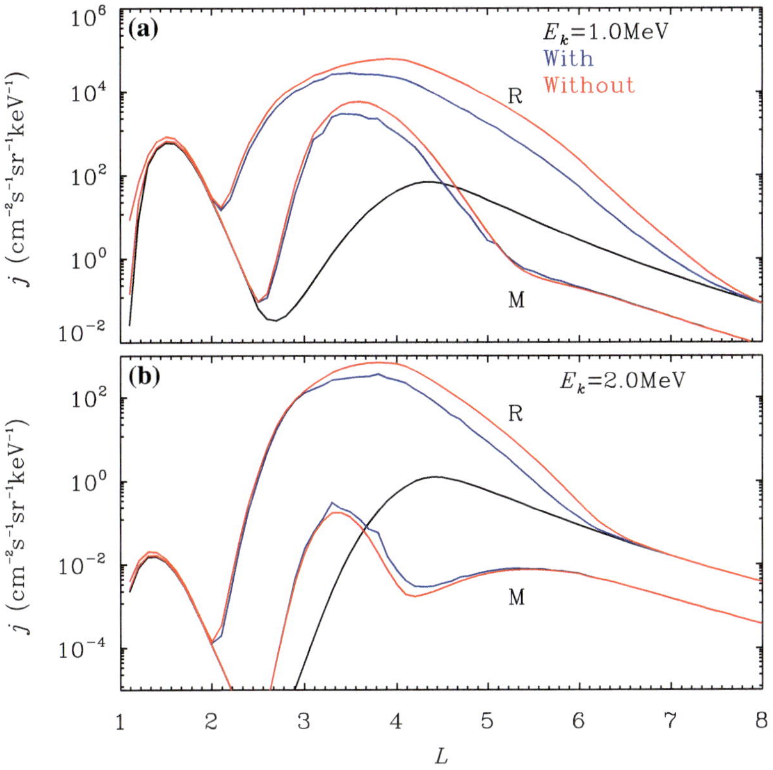

Fig. 3.11 The same as Fig. 3.9 except for electron fluxes at energies E_k = **a** 1.0 and **b** 2.0 MeV

We have implemented some idealized simulations and obtained the following results:

1. Some of the dominant characteristics of electron radiation belt during both quiet and active periods can be well reproduced by STEERB code. The obtained stable configuration, inner and outer radiation belts separated by slot region, generally comparable with the previous observations (Thorne et al. 2007). The energetic outer zone electron fluxes can decrease by one order or above during storm main phase, and substantially increase by several orders during storm recovery phase in the range $4 < L < 6$, comparable with previous observations (Reeves et al. 2003; Green et al. 2004) and numerical studies (Shprits et al. 2009). Observations (Reeves et al. 2003) also showed that ~50 % of storms do not produce significant enhancement of energetic electron fluxes in the heart of outer radiation belt. Such characteristics may be obtained by numerical model with relatively weaker chorus waves present during the recovery phase (Li et al. 2007). The electron PSD profiles with constant μ and J persistently peak in the range $4 < L < 6$ during storms, consistent with observations (Green and Kivelson 2004).

2. The respective roles of radial diffusion, cyclotron resonant interaction with chorus, and plume waves in the global radiation belt dynamics are evaluated. Numerical results support the previous findings that the dropouts of energetic electron fluxes during the main phase may be caused by combined cyclotron resonant interaction with plume waves and outward radial diffusion, while the buildups of energetic electron fluxes during the recovery phase can be produced by combined cyclotron resonant interaction with chorus waves and inward radial diffusion.
3. The importance of cross-diffusion in three-dimensional simulations is also investigated by comparison. The difference caused by ignoring cross-diffusion is relatively small during the main phase, while the equatorially-trapped energetic electron PSD can be overestimated by a factor of $5-10$ in the heart of outer radiation belt during the recovery phase, generally consistent with earlier studies (e.g., Su et al. 2011c). However, neglect of cross-diffusion may not distort the dominant characteristics of energetic electron PSD during both main and recovery phase especially at large pitch-angles. It should be noted that the precise difference between the simulations with and without cross-diffusion depends on a variety of factors, e.g., initial condition, boundary condition, and wave models, which should be parametrically studied to give a comprehensive assessment on the importance of cross-diffusion in future studies.

References

Abel B, Thorne RM (1998) Electron scattering loss in Earth's inner magnetosphere 1. Dominant physical processes. J Geophys Res 103:2385–2396. doi:10.1029/97JA02919

Albert JM, Young SL (2005) Multidimensional quasi-linear diffusion of radiation belt electrons. Geophys Res Lett 32(L14):110. doi:10.1029/2005GL023191

Albert JM, Meredith NP, Horne RB (2009) Three-dimensional diffusion simulation of outer radiation belt electrons during the october 9, 1990, magnetic storm. J Geophys Res 114(A09):214. doi:10.1029/2009JA014336

Beutier T, Boscher D (1995) A three-dimensional analysis of the electron radiation belt by the salammbô code. J Geophys Res 100:14853–14862. doi:10.1029/94JA03066

Bourdarie S, Boscher D, Beutier T, Sauvaud J, Blanc M (1997) Electron and proton radiation belt dynamic simulations during storm periods: a new asymmetric convection-diffusion model. J Geophys Res 102:17541–17552. doi:10.1029/97JA01305

Brautigam DH, Albert JM (2000) Radial diffusion analysis of outer radiation belt electrons during the October 9, 1990, magnetic storm. J Geophys Res 105:291–310. doi:10.1029/1999JA900344

Carpenter DL, Anderson RR (1992) An ISEE/whistler model of equatorial electron density in the magnetosphere. J Geophys Res 97:1097–1108

Christon SP, Mitchell DG, Williams DJ, Frank LA, Huang CY, Eastman TE (1988) Energy spectra of plasma sheet ions and electrons from about 50 eV/e to about 1 MeV during plamsa temperature transitions. J Geophys Res 93:2562–2572. doi:10.1029/JA093iA04p02562

Elkington SR, Hudson MK, Chan AA (1999) Acceleration of relativistic electrons via drift-resonant interaction with toroidal-mode Pc-5 ULF oscillations. Geophys Res Lett 26:3273–3276. doi:10.1029/1999GL003659

Elkington SR, Hudson MK, Chan AA (2003) Resonant acceleration and diffusion of outer zone electrons in an asymmetric geomagnetic field. J Geophys Res 108:1116. doi:10.1029/2001JA009202

Fei Y, Chan AA, Elkington SR, Wiltberger MJ (2006) Radial diffusion and MHD particle simulations of relativistic electron transport by ULF waves in the september 1998 storm. J Geophys Res 111(A12):209. doi:10.1029/2005JA011211

Fok MC, Horne RB, Meredith NP, Glauert SA (2008) Radiation belt environment model: application to space weather nowcasting. J Geophys Res 113(A12), A03S08. doi: 10.1029/2007JA012558

Green JC, Kivelson MG (2004) Relativistic electrons in the outer radiation belt: differentiating between acceleration mechanisms. J Geophys Res 109(A03):213. doi:10.1029/2003JA010153

Green JC, Onsager TG, O'Brien TP, Baker DN (2004) Testing loss mechanisms capable of rapidly depleting relativistic electron flux in the Earth's outer radiation belt. J Geophys Res 109(A12):211. doi:10.1029/2004JA010579

Horne RB, Thorne RM, Shprits YY, Meredith NP, Glauert SA, Smith AJ, Kanekal SG, Baker DN, Engebretson MJ, Posch JL, Spasojevic M, Inan US, Pickett JS, Decreau PME (2005) Wave acceleration of electrons in the Van Allen radiation belts. Nature 437:227–230. doi:10.1038/nature03939

Hudson MK, Elkington SR, Lyon JG, Goodrich CC (2000) Increase in relativistic electron flux in the inner magnetosphere: ULF wave mode structure. Adv Space Res 25:2327–2337. doi:10.1016/S0273-1177(99)00518-9

Jordanova VK, Miyoshi Y (2005) Relativistic model of ring current and radiation belt ions and electrons: initial results. Geophys Res Lett 32(L14):104. doi:10.1029/2005GL023020

Jordanova VK, Albert J, Miyoshi Y (2008) Relativistic electron precipitation by EMIC waves from self-consistent global simulations. J Geophys Res 113, A00A10. doi: 10.1029/2008JA013239

Komatsu K, Watanabe S (2008) Radial diffusion coefficient for the inner part of the Earth's electron radiation belt. Geophys Bull Hokkaido Univ 71:61–70

Li W, Shprits YY, Thorne RM (2007) Dynamic evolution of energetic outer zone electrons due to wave-particle interactions during storms. J Geophys Res 112(A11), A10,220. doi: 10.1029/2007JA012368

Li X, Temerin M, Baker DN, Reeves GD, Larson D (2001) Quantitative prediction of radiation belt electrons at geostationary orbit based on solar wind measurements. Geophys Res Lett 28: 1887–1890. doi:10.1029/2000GL012681

Li X, Barker AB, Baker DN, Tu WC, Sarris TE, Selesnick RS, Friedel R, Shen C (2009) Modeling the deep penetration of outer belt electrons during the "Halloween" magnetic storm in 2003. Space Weather 02:004. doi:10.1029/2008SW000418

Lyons LR, Thorne RM (1973) Equilibrium structure of radiation belt electrons. J Geophys Res 78:2142–2149. doi:10.1029/JA078i013p02142

Maksimovic M, Pierrard V, Lemaire JF (1997a) A kinetic model of the solar wind with Kappa distribution functions in the corona. Astron Astrophys 324:725–734

Maksimovic M, Pierrard V, Riley P (1997b) Ulysses electron distributions fitted with kappa functions. Geophys Res Lett 24:1151–1154. doi:10.1029/97GL00992

Meredith NP, Horne RB, Glauert SA, Anderson RR (2007) Slot region electron loss timescales due to plasmaspheric hiss and lightning-generated whistlers. J Geophys Res 112:8214. doi:10.1029/2007JA012413

Miyoshi YS, Jordanova VK, Morioka A, Thomsen MF, Reeves GD, Evans DS, GreenJC (2006) Observations and modeling of energetic electron dynamics during the October 2001 storm. J Geophys Res 111, A11S02. doi: 10.1029/2005JA011351

Reeves GD, McAdams KL, Friedel RHW, O'Brien TP (2003) Acceleration and loss of relativistic electrons during geomagnetic storms. Geophys Res Lett 30:1529. doi:10.1029/2002GL016513

Sarris T, Li X, Temerin M (2006) Simulating radial diffusion of energetic (MeV) electrons through a model of fluctuating electric and magnetic fields. Ann Geophys 24:2583–2598

Schulz M (1991) Geomagnetism, chap. The magnetosphere, geomagnetically trapped radiation. Elsevier, New York, pp 202–256

Schulz M, Lanzerotti LJ (1974) Particle diffusion in the radiation belts. Physics and chemistry in space, vol 7. Springer, New York

Sheeley BW, Moldwin MB, Rassoul HK, Anderson RR (2001) An empirical plasmasphere and trough density model: crres observations. J Geophys Res 106:25631–25642. doi:10.1029/2000JA000286
Shprits YY, Thorne RM (2004) Time dependent radial diffusion modeling of relativistic electrons with realistic loss rates. Geophys Res Lett 31(L08):805. doi:10.1029/2004GL019591
Shprits YY, Thorne RM, Friedel R, Reeves GD, Fennell J, Baker DN, Kanekal SG (2006a) Outward radial diffusion driven by losses at magnetopause. J Geophys Res 111(A11):214. doi:10.1029/2006JA011657
Shprits YY, Thorne RM, Horne RB, Glauert SA, Cartwright M, Russell CT, Baker DN, Kanekal SG (2006b) Acceleration mechanism responsible for the formation of the new radiation belt during the 2003 halloween solar storm. Geophys Res Lett 33(L05):104. doi:10.1029/2005GL024256
Shprits YY, Subbotin DA, Meredith NP, Elkington SR (2008) Review of modeling of losses and sources of relativistic electrons in the outer radiation belt II: local acceleration and loss. J Atmos Sol Terr Phys 70:1694–1713. doi:10.1016/j.jastp.2008.06.014
Shprits YY, Chen L, Thorne RM (2009a) Simulations of pitch angle scattering of relativistic electrons with MLT-dependent diffusion coefficients. J Geophys Res 114(A03):219. doi:10.1029/2008JA013695
Shprits YY, Subbotin D, Ni B (2009b) Evolution of electron fluxes in the outer radiation belt computed with the VERB code. J Geophys Res 114(A11):209. doi:10.1029/2008JA013784
Strang G (1968) On the construction and comparison of difference schemes. SIAM J Numer Anal 5:506
Su Z, Zheng H, Wang S (2009a) Dynamic evolution of energetic outer zone electrons due to whistler-mode chorus based on a realistic density model. J Geophys Res 114(A07):201. doi:10.1029/2008JA014013
Su Z, Zheng H, Wang S (2009b) Evolution of electron pitch angle distribution due to interactions with whistler-mode chorus following substorm injections. J Geophys Res 114(A08):202. doi:10.1029/2009JA014269
Su Z, Zheng H, Xiong M (2009c) Dynamic evolution of outer radiation belt electrons due to whistler-mode chorus. Chin Phys Lett 26(039):401
Su Z, Xiao F, Zheng H, Wang S (2010a) STEERB: a three-dimensional code for storm-time evolution of electron radiation belt. J Geophys Res 115(A09):208. doi:10.1029/2009JA015210
Su Z, Xiao F, Zheng H, Wang S (2010b) Combined radial diffusion and adiabatic transport of radiation belt electrons with arbitrary pitch-angles. J Geophys Res 115(A10):249. doi:10.1029/2010JA015903
Su Z, Zheng H, Wang S (2010c) A parametric study on the diffuse auroral precipitation by resonant interaction with whistler-mode chorus. J Geophys Res 115(A05):219. doi:10.1029/2009JA014759
Su Z, Zheng H, Wang S (2010d) Three dimensional simulations of energetic outer zone electron dynamics due to wave-particle interaction and azimuthal advection. J Geophys Res 115(A06):203. doi:10.1029/2009JA014980
Su Z, Xiao F, Zheng H, Wang S (2011a) CRRES observation and STEERB simulation of the 9 October 1990 electron radiation belt dropout event. Geophys Res Lett 38(L06):106. doi:10.1029/2011GL046873
Su Z, Xiao F, Zheng H, Wang S (2011b) Radiation belt electron dynamics driven by adiabatic transport, radial diffusion, and wave-particle interactions. J Geophys Res 116(A04):205. doi:10.1029/2010JA016228
Su Z, Zheng H, Chen L, Wang S (2011c) Numerical simulations of storm-time outer radiation belt dynamics by wave-particle interactions including cross diffusion. J Atmos Sol Terr Phys 73:95–105. doi:10.1016/j.jastp.2009.08.002
Subbotin DA, Shprits YY (2009) Three-dimensional modeling of the radiation belts using the versatile electron radiation belt (VERB) code. Space Weather 10:001. doi:10.1029/2008SW000452
Tao X, Chan AA, Albert JM, Miller JA (2008) Stochastic modeling of multidimensional diffusion in the radiation belts. J Geophys Res 113(A07):212. doi:10.1029/2007JA012985

Tao X, Albert JM, Chan AA (2009) Numerical modeling of multidimensional diffusion in the radiation belts using layer methods. J Geophys Res 114(A02):215. doi:10.1029/2008JA013826

Thorne RM, O'Brien TP, Shprits YY, Summers D, Horne RB (2005) Timescale for MeV electron microburst loss during geomagnetic storms. J Geophys Res 110(A09):202. doi:10.1029/2004JA010882

Thorne RM, Shprits YY, Meredith NP, Horne RB, Li W, Lyons LR (2007) Refilling of the slot region between the inner and outer electron radiation belts during geomagnetic storms. J Geophys Res 112(A06):203. doi:10.1029/2006JA012176

Varotsou A, Boscher D, Bourdarie S, Horne RB, Glauert SA, Meredith NP (2005) Simulation of the outer radiation belt electrons near geosynchronous orbit including both radial diffusion and resonant interaction with whistler-mode chorus waves. Geophys Res Lett 32(L19):106. doi:10.1029/2005GL023282

Varotsou A, Boscher D, Bourdarie S, Horne RB, Meredith NP, Glauert SA, Friedel RH (2008) Three-dimensional test simulations of the outer radiation belt electron dynamics including electron-chorus resonant interactions. J Geophys Res 113(A12):212. doi:10.1029/2007JA012862

Vasyliunas VM (1968) A survey of low-energy electrons in the evening sector of the magnetosphere with OGO 1 and OGO 3. J Geophys Res 73:2839

Viñas AF, Mace RL, Benson RF (2005) Dispersion characteristics for plasma resonances of Maxwellian and Kappa distribution plasmas and their comparisons to the IMAGE/RPI observations. J Geophys Res 110(A06):202. doi:10.1029/2004JA010967

Wentworth RC, MacDonald WM, Singer SF (1959) Lifetimes of trapped radiation belt particles determined by coulomb scattering. Phys Fluids 2:499–509. doi:10.1063/1.1705940

Xiao F (2006) Modelling energetic particles by a relativistic kappa loss cone distribution function in plasmas. Plasma Phys Control Fusion 48:203

Xiao F, Chen L, Li J (2008a) Energetic particles modeled by a generalized relativistic kappa-type distribution function in plasmas. Plasma Phys Control Fusion 50(105):002. doi:10.1088/0741-3335/50/10/105002

Xiao F, Shen C, Wang Y, Zheng H, Wang S (2008b) Energetic electron distributions fitted with a relativistic kappa-type function at geosynchronous orbit. J Geophys Res 113(A05):203. doi:10.1029/2007JA012903

Xiao F, Zhou Q, Li C, Cai A (2008c) Modeling solar energetic particle by a relativistic kappa-type distribution. Plasma Phys Control Fusion 50(062):001. doi:10.1088/0741-3335/50/6/062001

Xiao F, Su Z, Zheng H, Wang S (2009) Modeling of outer radiation belt electrons by multidimensional diffusion process. J Geophys Res 114(A03):201. doi:10.1029/2008JA013580

Xiao F, Su Z, Zheng H, Wang S (2010) Three-dimensional simulations of outer radiation belt electron dynamics including cross diffusion terms. J Geophys Res 115(A05):216. doi:10.1029/2009JA014541

Chapter 4
Adiabatic Transport

4.1 Introduction

Trapped particles in the inner magnetosphere have three adiabatic invariants associated to the three periodic motions (see Sect. 1.3.1). Consequently, the physical processes controlling radiation belt evolution can be classified into adiabatic and nonadiabatic processes. Two nonadiabatic processes, cyclotron resonance, and radial diffusion, have been investigated in the Chaps. 2 and 3. In fact, during the geomagnetic storm, the geomagnetic field configuration can change significantly, as a result of the buildup or decay of ring current (Dessler and Karplus 1961; McIlwain 1966; Hilmer and Voigt 1995; Tsyganenko 1995; Ganushkina et al. 2002). In response to such large-scale variation in geomagnetic field, the trapped electrons could experience the adiabatic transport process (conserving the three adiabatic invariants but changing the pitch-angle, energy, and drift-shell L). During the main phase of a storm, the buildup of ring current can reduce the magnetic field strength. To conserve the third adiabatic invariant, the electron should move toward larger L region. The corresponding kinetic energy needs to decrease with the decrease of magnetic field in order to conserve the first the adiabatic invariant. During the recovery phase, a reversible process occur as a result of the decay of ring current. In this chapter, we first improve the global electron radiation belt diffusion model STEERB by including the adiabatic transport process, next give some idealized the simulations to investigate the effect of adiabatic transport on radiation belt evolution, and at last simulate the electron radiation belt dropout event on 9 October 1990 to identify the dominant loss mechanisms.

Z. Su, *A Global Kinetic Model for Electron Radiation Belt Formation and Evolution*, Springer Theses, DOI 10.1007/978-3-662-46651-3_4

4.2 Improved Global Radiation Belt Diffusion Model STEERB

4.2.1 Background Magnetic Field

The *Dst*-dependent symmetrical Hilmer-Voigt geomagnetic field model (Hilmer and Voigt 1995) is adopted to determine the mapping relation between the adiabatic invariants and observable variables. In the spherical coordinate system (r, θ, ϕ), the geomagnetic field can be expressed as

$$\mathbf{B} = \nabla \times \mathbf{A} = \nabla \times (A_\phi \mathbf{e}_\phi), \tag{4.1}$$

$$A_\phi = \left[-\frac{B_0}{r^2} + \frac{4B_+ r_+^3 r}{(r^2 + 4r_+^2)^{3/2}} + \frac{4B_- r_-^3 r}{(r^2 + 4r_-^2)^{3/2}} \right] \sin\theta. \tag{4.2}$$

Here the ϕ-component of vector potential (4.2) consists of the contributions of the intrinsic dipole magnetic field (first term) with the equatorial field magnitude at the Earths surface B_0, and the external magnetic field related to the ring current with four *Dst*-dependent parameters r_+, r_-, B_+ and B_- (see Kim and Chan 1997; Su et al. 2010b). It should be mentioned that, to save computing time, the dipole geomagnetic field model is used for the local diffusion driven by cyclotron resonance.

4.2.2 Basic Equation

The evolution of radiation belt electron PSD f can be described by the following three-dimensional diffusion equation (e.g., Schulz and Lanzerotti 1974; Su et al. 2010b, 2011a, b)

$$\begin{aligned} \frac{\partial f}{\partial t} = & \frac{1}{Gp} \frac{\partial}{\partial \alpha_e}\bigg|_{p,L} \left[G \left(\langle D_{\alpha\alpha} \rangle \frac{1}{p} \frac{\partial f}{\partial \alpha_e}\bigg|_{p,L} + \langle D_{\alpha p} \rangle \frac{\partial f}{\partial p}\bigg|_{\alpha_e,L} \right) \right] \\ & + \frac{1}{G} \frac{\partial}{\partial p}\bigg|_{\alpha_e,L} \left[G \left(\langle D_{p\alpha} \rangle \frac{1}{p} \frac{\partial f}{\partial \alpha_e}\bigg|_{p,L} + \langle D_{pp} \rangle \frac{\partial f}{\partial p}\bigg|_{\alpha_e,L} \right) \right] \\ & + L^{*2} \frac{\partial}{\partial L^*}\bigg|_{\mu,J} \left(\frac{D_{L^*L^*}}{L^{*2}} \frac{\partial f}{\partial L^*}\bigg|_{\mu,J} \right) - \frac{f}{\tau_L} - \frac{f}{\tau_C} - \frac{f}{\tau_W}. \end{aligned} \tag{4.3}$$

The present STEERB model involves two computational spaces: Adiabatic invariant space (μ, J, L^*) and observable space (α_e, p, L). The mapping from adiabatic invariant space to observable space can self-consistently incorporate the adiabatic transport related to magnetic field variation. The first two terms on the right-hand side of Eq. (4.3) treat the local diffusion processes with the drift-averaged diffusion coefficients in pitch-angle $\langle D_{\alpha\alpha}\rangle$, momentum $\langle D_{pp}\rangle$, and cross-pitch-angle-momentum $\langle D_{\alpha p}\rangle = \langle D_{p\alpha}\rangle$ (see Chap. 2). The third term on the right-hand side of Eq. (4.3) treats the radial diffusion process with the radial diffusion coefficient $D_{L^*L^*}$ (see Chap. 3). The last three terms (Su et al. 2010a, b) present the dominant loss mechanisms in the loss cone, near the Earth ($L < 1.5$), and in the slot region, respectively (see Chap. 3).

The mapping relation between the two computational spaces is defined as

$$\mu = \frac{p^2 \sin^2 \alpha_e}{2m_e B_e}, \tag{4.4a}$$

$$J = 2pI = 2p \int_{s_m}^{s_m'} \sqrt{1 - \frac{B}{B_e} \sin^2 \alpha_e} ds, \tag{4.4b}$$

$$L^* \equiv \frac{2\pi R_E^2 B_0}{\oint \mathbf{A} \cdot dl} = \frac{1}{L} - 4\frac{B_+}{B_0} \frac{r_+^3 L^2}{(L^2 + 4r_+^2)^{\frac{3}{2}}} - 4\frac{B_-}{B_0} \frac{r_-^3 L^2}{(L^2 + 4r_-^2)^{\frac{3}{2}}}. \tag{4.4c}$$

Here B_e is the equatorial geomagnetic field, and other parameters and variables have been specified in the Sect. 1.3.1. Figure 4.1 shows several examples of the geomagnetic field configurations and sample relations between adiabatic invariants and observables with different *Dst* values.

4.2.3 Diffusion Coefficients

At a fixed L^*, the local diffusion coefficients are the same as those in the Chap. 3. The radial diffusion coefficient is written as (Brautigam and Albert 2000; Albert et al. 2009)

$$D_{L^*L^*} = P_M D_{L^*L^*}^M + P_E D_{L^*L^*}^E, \tag{4.5}$$

$$D_{L^*L^*}^M = 10^{0.506K_p - 9.325} L^{*10} \quad [1/\mathrm{d}], \tag{4.6}$$

$$D_{L^*L^*}^E = \frac{c^2 \widetilde{E}^2}{4R_E^2 B_0^2} \frac{T_d}{1 + (\omega_d T_d/2)^2} L^{*6}, \tag{4.7}$$

Two physical parameters $\widetilde{E}$ and T_d have been listed in the Chap. 3, and two tunable parameter P_M and P_E need to be determined in the following sections.

4.2.4 Numerical Method

Based on the split-operator technique, we solve the local diffusion part by the HFD scheme (Su et al. 2009a, b, c, 2010c, d, 2011c; Xiao et al. 2009, 2010), and solve

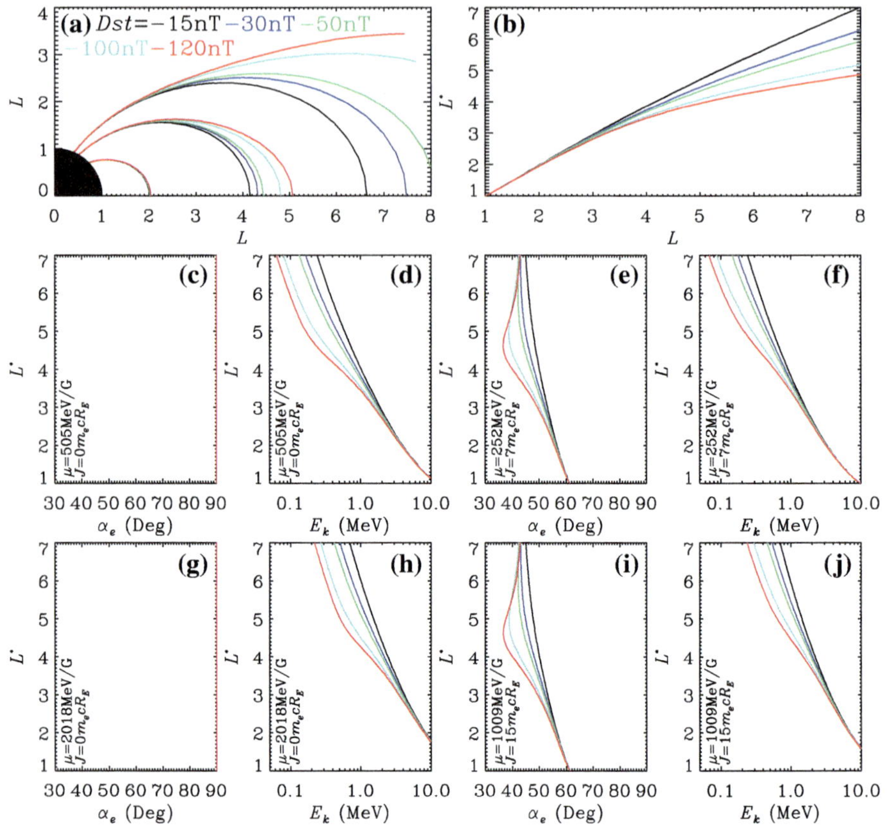

Fig. 4.1 **a** Geomagnetic field lines in the meridian plane, **b–j** Sample relations between adiabatic invariants (μ, J, L^*) and observables (α_e, E_k, L). The *solid*, *dotted*, *dashed*, *dot-dashed* and *dot-dot-dashed lines* represent the simulation cases with $Dst = -15$, -30, -50, -100 and -120 nT, respectively

the radial diffusion part by the fully implicit scheme. The mapping between two computational spaces is implemented by the spline interpolation method.

The boundary conditions are set to be

$$\frac{\partial f}{\partial \alpha_e}(\alpha_e = 0°) = 0, \qquad \frac{\partial f}{\partial \alpha_e}(\alpha_e = 90°) = 0, \tag{4.8}$$

$$f(E_k = 0.1\,\text{MeV}) = f_l(t, \alpha_e, L), \qquad f(E_k = 5\,\text{MeV}) = f_u(t, \alpha_e, L), \tag{4.9}$$

$$f(L^* = 1) = 0, \qquad f(L^* = 7) = f_o(t, \mu, J), \tag{4.10}$$

where $f_l(t, \alpha_e, L)$ and $f_u(t, \alpha_e, L)$ are interpolated from the PSD in the adiabatic invariant space at each step, and $f_o(t, \mu, J)$ is determined based on observations or empirical relations.

4.3 Idealized Simulations

4.3.1 Fully Adiabatic Transport

In this section, the initial distribution is assumed to be the steady-state solution of the radial diffusion equation

$$L^{*2}\left.\frac{\partial}{\partial L^*}\right|_{\mu,J}\left(\frac{D_{L^*L^*}}{L^{*2}}\left.\frac{\partial f}{\partial L^*}\right|_{\mu,J}\right)-\frac{f}{\tau_C}-\frac{f}{\tau_W}-\frac{f}{\tau_E}=0. \tag{4.11}$$

It should be mentioned that the Eq. (4.11) includes an additional loss term $-f/\tau_E$. τ_E represent the lifetime of electrons out of plasmapause (Shprits et al. 2006)

$$\tau_E = 6/K_p \quad (L > L_{pp}) \quad [\text{d}]. \tag{4.12}$$

The geomagnetic indices are set to be $K_p = 2$ and $Dst = -15$ nT (corresponding to the quiet time). The outer boundary condition is determined as

$$f_o(t=0,\mu,J) = F(E_k)\sin^q\alpha_e, \tag{4.13}$$

where $F(E_k) = 8222.6\exp(-7.068E_k)/p^2$ is an average PSD function obtained from CRRES and Polar measurements (Shprits et al. 2006), and $n = 0.852$ is the pitch-angle index at $L^* = 7$ ($L \approx 8$ when $Dst = -15$ nT) (Thorne et al. 2005).

Analogous to the Sect. 3.3.1, the strength of radial diffusion is adjusted in order to reproduce the electron flux profiles observed by MEA instrument at Orbit 182 of CRRES satellite. The optimal parameters in radial diffusion coefficients would be

$$P_m = 1.0, \tag{4.14}$$

$$P_e = \begin{cases} 1.0, & \text{when } L < 2.2; \\ 3.2 - L, & \text{when } 2.2 < L < 3.0; \\ 0.2, & \text{when } L > 3.0. \end{cases} \tag{4.15}$$

The obtained steady-state fluxes $j = p^2 f$ of equatorially-trapped electrons in the observable space are shown in Fig. 4.2. Clearly, the modeling flux profiles are generally consistent with observations especially in the outer radiation belt.

In this section, we simulate the fully adiabatic transport of radiation belt electrons for an idealized storm. The Dst index is assumed to be

$$Dst = \begin{cases} -105t - 15, & 0.0 \le t < 1.0\text{ d} \\ 35t - 155, & 1.0 \le t < 4.0\text{ d} \\ -15, & t \ge 4.0\text{ d} \end{cases}. \tag{4.16}$$

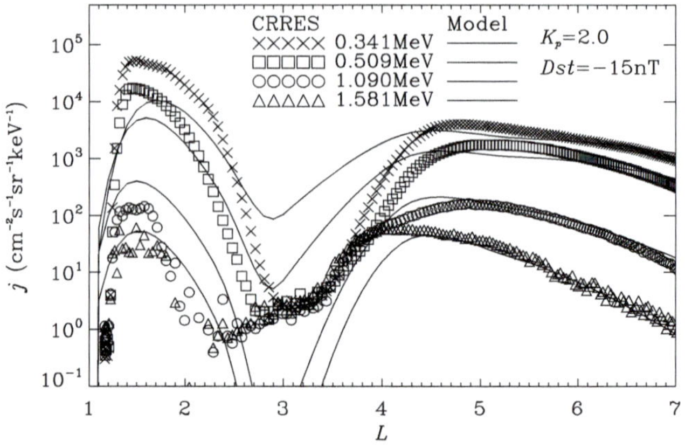

Fig. 4.2 *Solid lines* Steady-state flux profiles of equatorially-trapped electrons with different kinetic energies. *Symbols* The corresponding electron flux profiles observed by MEA instrument at Orbit 182 of CRRES satellite

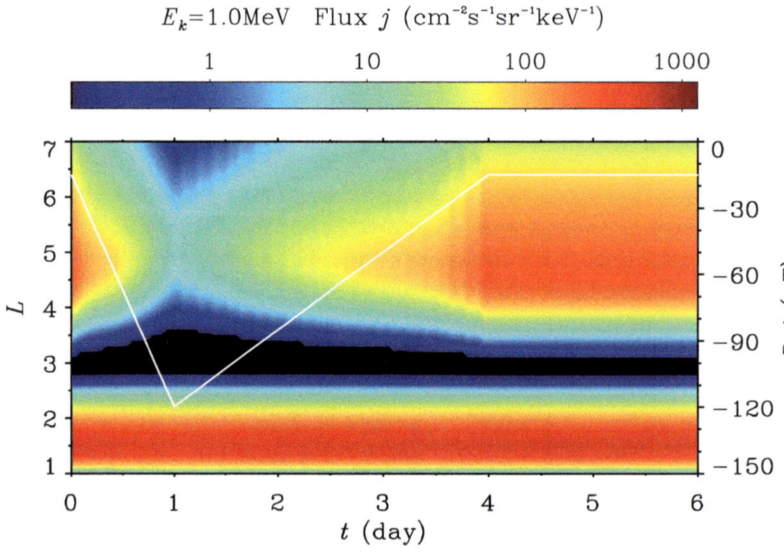

Fig. 4.3 Adiabatic evolution of equatorially-trapped 1 MeV electron fluxes $j = p^2 f$. Note that the *white line* denotes the Dst index

Dst has the maximum −15 nT during the quiet time, and the minimum −120 nT during the main phase.

The simulated adiabatic evolution of equatorially-trapped 1 MeV electron fluxes j is plotted in the Fig. 4.3, and the corresponding radial profiles of electron fluxes at the quite time ($Dst = -15$ nT) and the Dst-minimum time ($Dst = -120$ nT) are shown in the Fig. 4.4. Clearly, the adiabatic transport can lead to the reversible change

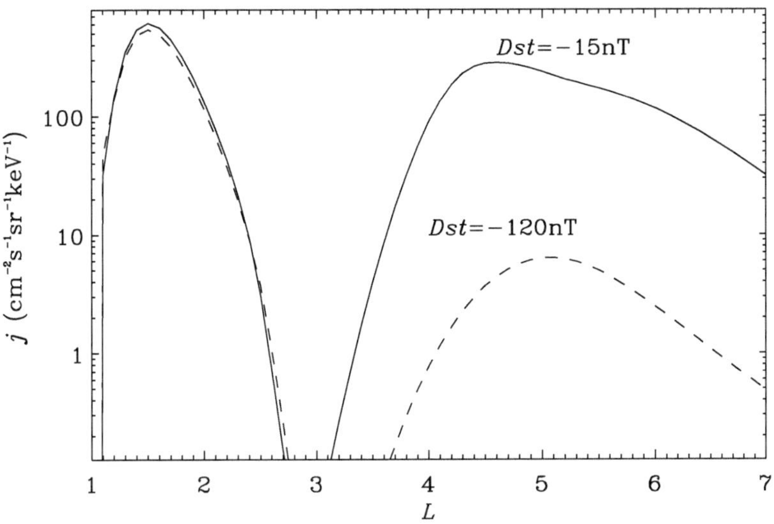

Fig. 4.4 Radial profiles of equatorially-trapped fluxes $j = p^2 f$ at the quite time ($Dst = -15\,\text{nT}$) and the Dst-minimum time ($Dst = -120\,\text{nT}$)

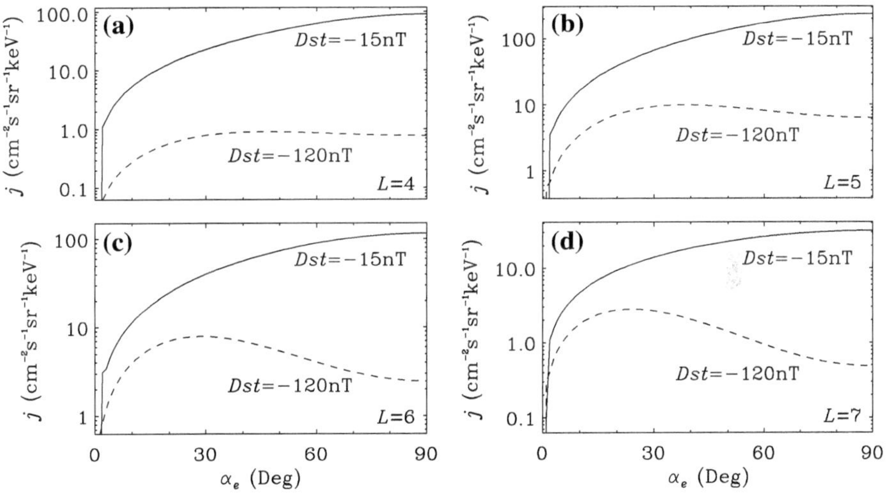

Fig. 4.5 Pitch-angle distributions of 1 MeV electrons at the quite time ($Dst = -15\,\text{nT}$) and the Dst-minimum time ($Dst = -120\,\text{nT}$)

of electron flux. The electron flux decreases by a factor of above 60 throughout the outer belt at $t = 1.0$ day, and returns to the prestorm value at $t = 4.0$ days.

Figure 4.5 presents the pitch-angle distribution of 1 MeV electrons during the quite time and the Dst-minimum time at $L = 4$, 5, 6 and 7. During the quiet time, the electron pitch-angle distributions are always rounded 90°-peaked. With the decrease of Dst index, the pitch-angle distributions gradually evolve into the so-called butterfly-shaped distributions, which is caused by the pitch-angle-dependent energy change of electrons for adiabatic transport in the current model. Such change in pitch-angle distributions would affect the local diffusion driven by cyclotron resonance.

4.3.2 Combination of Adiabatic and Nonadiabatic Processes

The initial condition is given by solving the steady-state radial diffusion equation with $K_p = 2.0$ and $Dst = -15\,\mathrm{nT}$

$$L^{*2} \frac{\partial}{\partial L^*}\bigg|_{\mu,J} \left(\frac{D_{L^*L^*}}{L^{*2}} \frac{\partial f}{\partial L^*}\bigg|_{\mu,J} \right) - \frac{f}{\tau_C} - \frac{f}{\tau_W} = 0. \tag{4.17}$$

The outer boundary condition is specified as

$$f_o(t=0,\mu,J) = f^\kappa(\alpha_e, E_k) \equiv c_\kappa \left[1 + \frac{E_k}{\kappa E_0} \right]^{-(\kappa+1)} \sin^q \alpha_e, \tag{4.18}$$

with $E_0 = 3$ keV, $\kappa = 5$, $q = 0.855$, $P_M = 1.0$, and $P_E = 0.2$. Figure 4.6 shows the profiles of obtained steady-state radiation belt electron fluxes $j = p^2 f$ in the observable space. The L-dependent configuration, inner and outer belts separated by slot region (2–3 R_E), can be clearly distinguished, generally comparable to the previous observations (e.g., Thorne et al. 2007). The pitch-angle anisotropy of ~MeV electrons is found to increase with L decreasing, consistent with the earlier measurements (see Thorne et al. 2005).

The current model mainly contains these physical processes: adiabatic transport (AT), radial diffusion (RD), cyclotron resonant interaction with chorus waves (CW), and plume hiss and EMIC waves (PW). Four simulation cases are conducted for an idealized storm: (A1) RD+CW; (A2) AT+RD+CW; (B1) RD+CW+PW; (B2) AT+RD+CW+PW. These simulation cases are further categorized into two groups: (A) = {A1, A2} (without plume waves) and (B) = {B1, B2} (with plume waves). The suffixes 1 and 2 denote the simulation cases without and with adiabatic transport. The turn on or off of adiabatic transport process is implemented by adopting a time-varying or static geomagnetic field model. The chorus, plume hiss, and EMIC waves are assumed to be distributed from $L = \max(3, L_{pp})$ to 7, where L_{pp} is the plasmapause location. These wave models are the same as those adopted in the Chap. 3. The evolution of Dst and K_p indices is described by Eqs. (4.16) and (3.18), and the time-dependent outer boundary conditions are assumed to be

$$f_o(t,\mu,J) = \overline{f}(t) f_o(t=0,\mu,J), \tag{4.19}$$

with $\overline{f}(t)$ given in the Sect. 3.3.2.

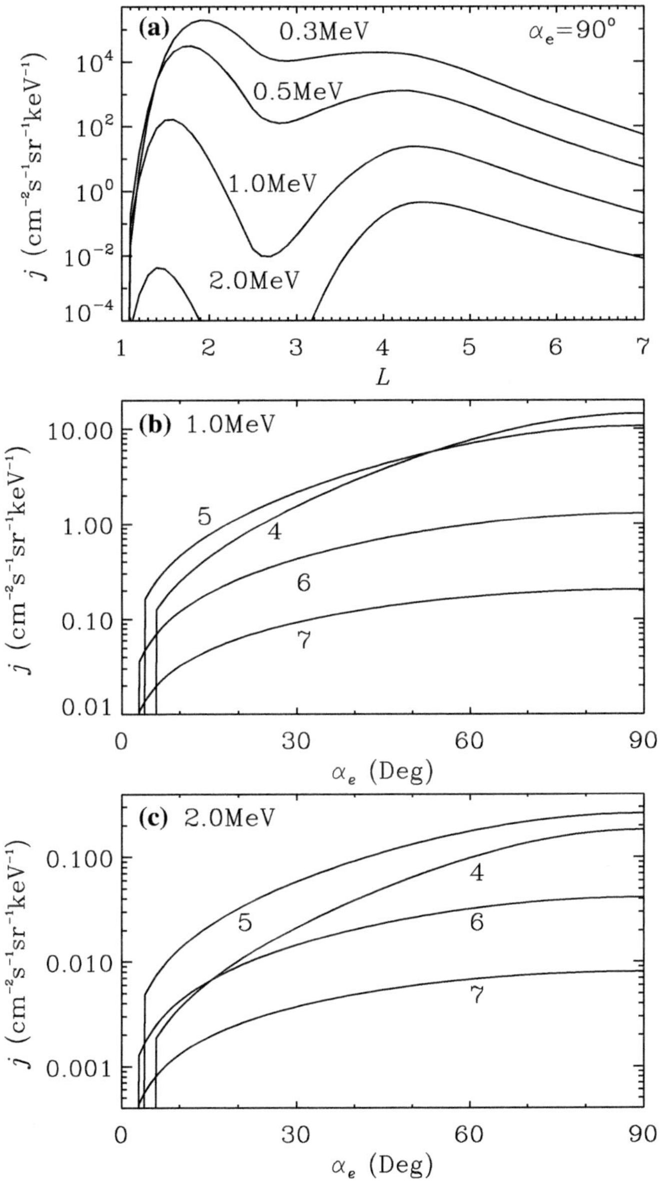

Fig. 4.6 **a** L-dependent flux profiles of equatorially-trapped electrons with different kinetic energies (shown). Pitch-angle-dependent flux profiles of 1.0 (**b**) and 2.0 MeV (**c**) electrons at different L-shells (shown)

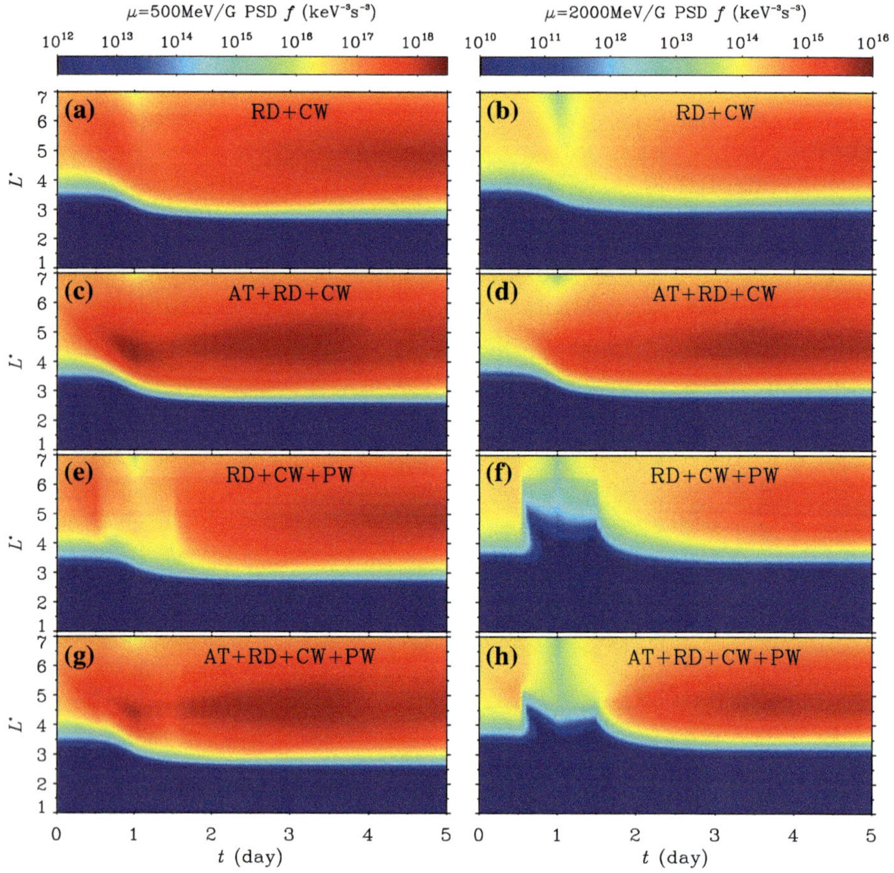

Fig. 4.7 Evolution of equatorially-trapped electron PSD at μ =500 (*left*) and 2000 MeV/G (*right*) in the adiabatic invariant space with **a**, **b** for Case A1, **c**, **d** for Case A2, **e**, **f** for Case B1, and **g**, **h** for Case B2

Evolutions of equatorially-trapped electron PSDs at μ = 500 and 2000 MeV/G for the four cases are shown in Fig. 4.7. The corresponding electron PSD profiles during the main phase (t = 1.0 d) and recovery phase (t = 4.0 d) are plotted in Fig. 4.8. For all the four simulation cases, the energetic electron PSDs exhibit significant (20–50 times) enhancements after storm in the region $4 < L^* < 6$, primarily due to the chorus-driven acceleration (e.g., Albert et al. 2009; Shprits et al. 2009; Su et al. 2010a). The comparison between Groups A (without plume waves) and B (with plume waves) indicates that incorporation of the plume wave-particle interaction can effectively deplete the PSDs especially for the high-energy electrons during the main phase, consistent with the previous numerical studies (e.g., Li et al. 2007; Shprits et al. 2009; Su et al. 2010a). The comparison between simulation cases in each group shows that ignoring of adiabatic transport process generally leads to

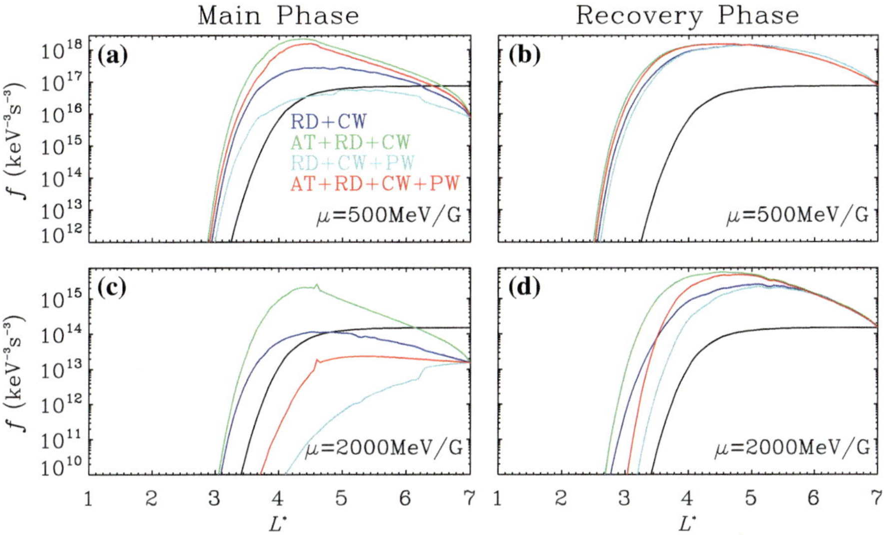

Fig. 4.8 Equatorially-trapped electron PSD profiles at μ = 500 (*up*) and 2000 MeV/G (*down*) in the adiabatic invariant space during main phase (*left*) and recovery phase (*right*) for Cases A1 (*dotted*), A2 (*dashed*), B1 (*dot-dashed*), and B2 (*dot-dot-dashed*). The *solid lines* represent the steady state

the underestimate of energetic electron PSDs throughout the storm. During the main phase (t = 1.0 d), the PSD underestimate factors at μ = 500 and 2000 MeV/G for both Group A and B are about 20–100 in the heart of outer radiation belt. Note that the kinetic energies of electrons possessing the constant adiabatic invariants decrease with Dst decreasing in the simulation Cases A2 and B2 (see Fig. 4.1). During the recovery phase (t = 4.0 d), the PSD underestimate factor at μ = 500 and 2000 MeV/G are about 2 and 5 for each group in the center of outer radiation belt. According to Liouville's theorem, the fully adiabatic transport alone (without other nonadiabatic processes) does not change the PSD function in the adiabatic invariant space. The current simulation results indicate that the adiabatic and nonadiabatic processes are coupled to each other, and that adiabatic transport, together with other nonadiabatic processes, can obviously affect the PSD function.

Figure 4.9 shows the evolutions of equatorially-trapped electron fluxes $j = p^2 f$ at E_k = 1.0 and 2.0 MeV for the four cases. Figure 4.10 plots the corresponding electron flux profiles during the main phase (t = 1.0 d) and recovery phase (t = 4.0 d). In the simulation (Case B2) including the adiabatic transport, radial diffusion, cyclotron resonant interaction with chorus and plume waves, the energetic outer radiation belt electron fluxes are found to decrease by about 2–3 orders of magnitude during the main phase, and substantially increase by about 2 orders of magnitude during the recovery phase, comparable to the previous observations (Reeves et al. 2003; Green et al. 2004). For the two cases of Group A (without plume waves), only the insignificant main phase depletion (less than 10 times) can occur beyond $5R_E$, primarily due to the outward radial diffusion induced by outer

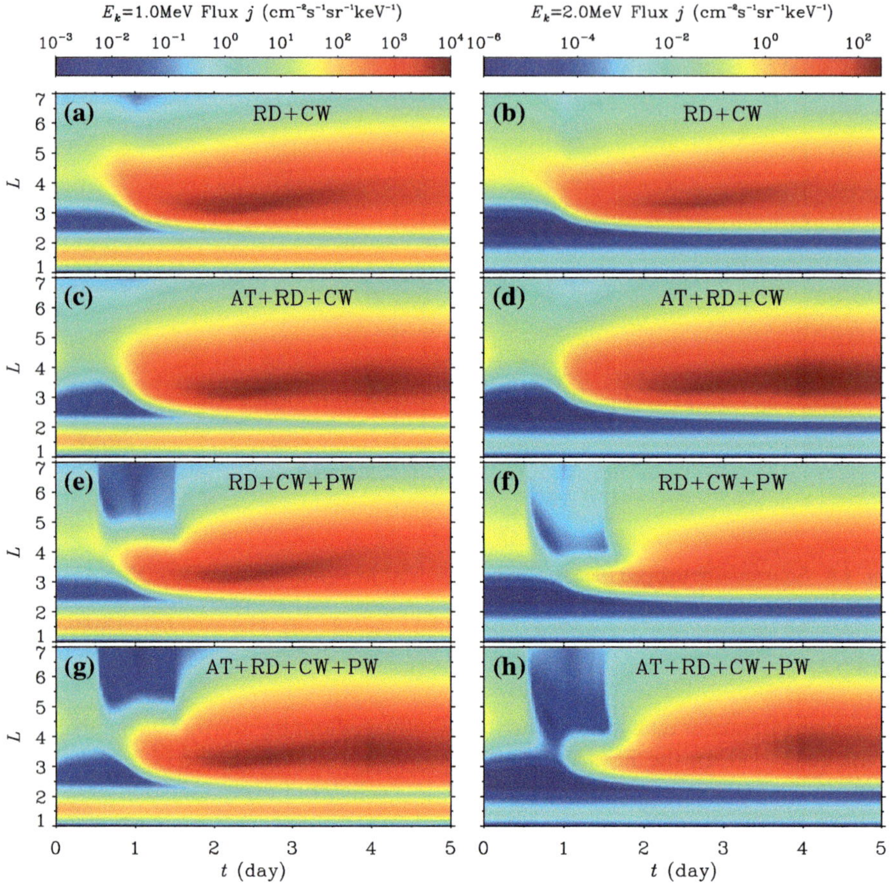

Fig. 4.9 The same as Fig. 4.7 except for electron fluxes j at energies $E_k = 1.0$ (*left*) and 2.0 MeV (*right*) in the observable space

boundary loss. The significant main phase depletion can be produced when the plume wave-particle interactions are included (Group B). These results suggest that adiabatic transport alone (without plume waves) is difficult to cause the significant depletion of energetic electron fluxes in the presence of strong chorus wave-particle interactions during the main phase. The comparison between simulation cases in each group shows that ignoring of adiabatic transport process generally yields an overestimate (underestimate) of energetic outer radiation belt electron fluxes during the main (recovery) phase. Such main phase overestimate in the Group B (with plume waves) is much stronger than that in the Group A (without plume waves). For the Group A, the main phase overestimate factors at energies 1.0 and 2.0 MeV are about 5 in the region $3 < L < 4$. For the Group B (with plume waves), the main phase overestimate factor at 2.0 MeV can approach ~30 in the region $4 < L < 5$, much larger than that at 1.0 MeV. The Groups A and B generally have the same recovery

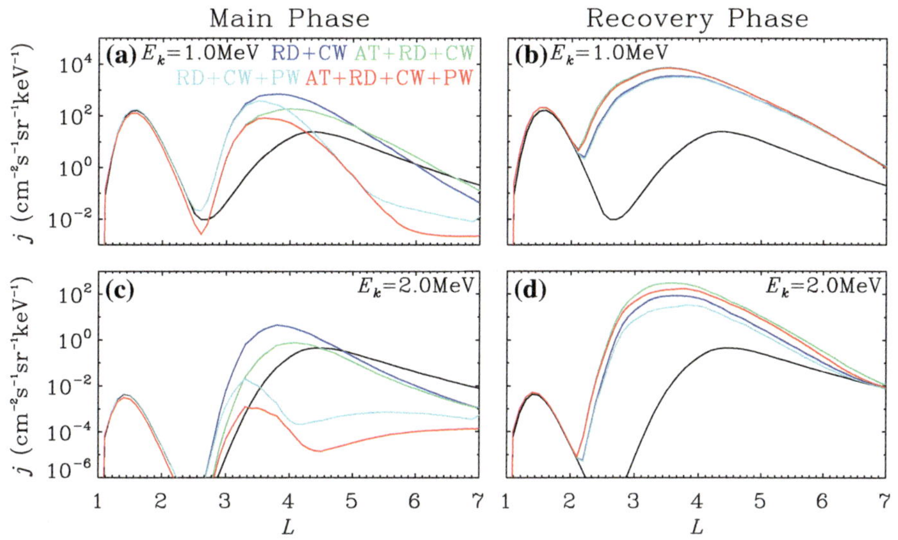

Fig. 4.10 The same as Fig. 4.8 except for electron fluxes j at energies $E_k = 1.0$ (*up*) and 2.0 MeV (*down*) in the observable space. In the panel (**b**), the *dotted* and *dot-dashed lines* overlap each other, and the *dashed* and *dot-dot-dashed lines* overlap each other

phase underestimate factor, which is relatively small ($\leq$2) for 1.0 MeV electron flux in the outer radiation belt region, and can approach 2–5 for 2.0 MeV electron flux in the region $3 < L < 6$.

The pitch-angle distributions of 1.0 and 2.0 MeV electrons for the four cases at different L-shells during the main phase ($t = 1.0$ d) and recovery phase ($t = 4.0$ d) are plotted in Figs. 4.11 and 4.12. All of the pitch-angle distributions (except the Case A2 at 2.0 MeV during the main phase) are 90°-peaked throughout the storm time. Our previous work (Su et al. 2010b) showed that combined radial diffusion and adiabatic transport can lead to the pitch-angle distribution evolution from a rounded 90°-peaked (butterfly-shaped) distribution to a butterfly-shaped (90°-peaked) distribution during the main (recovery) phase in the outer radiation belt. The persistence of 90°-peaked distribution in the current simulations may be caused by the pitch-angle scattering (more effective at small pitch-angles) and energization (more effective at large pitch-angles) related to various wave-particle interactions. The strong enhancement of energetic electron fluxes over all the pitch-angles during the recovery phase is found to occur in the four simulation cases. Only when the cyclotron resonant interaction with plume waves is introduced in the numerical model (Group B), the main phase loss of energetic electron fluxes over all the pitch-angles can form. Due to the neglect of adiabatic response, the main phase overestimate of energetic electron fluxes can occur over all the pitch-angles, while the recovery phase underestimate of energetic electron fluxes mainly occurs at the relatively large pitch-angles ($\alpha_e > 50°$). Clearly, the storm time overestimate or underestimate is more significant at the energy channel 2.0 MeV than that at 1.0 MeV.

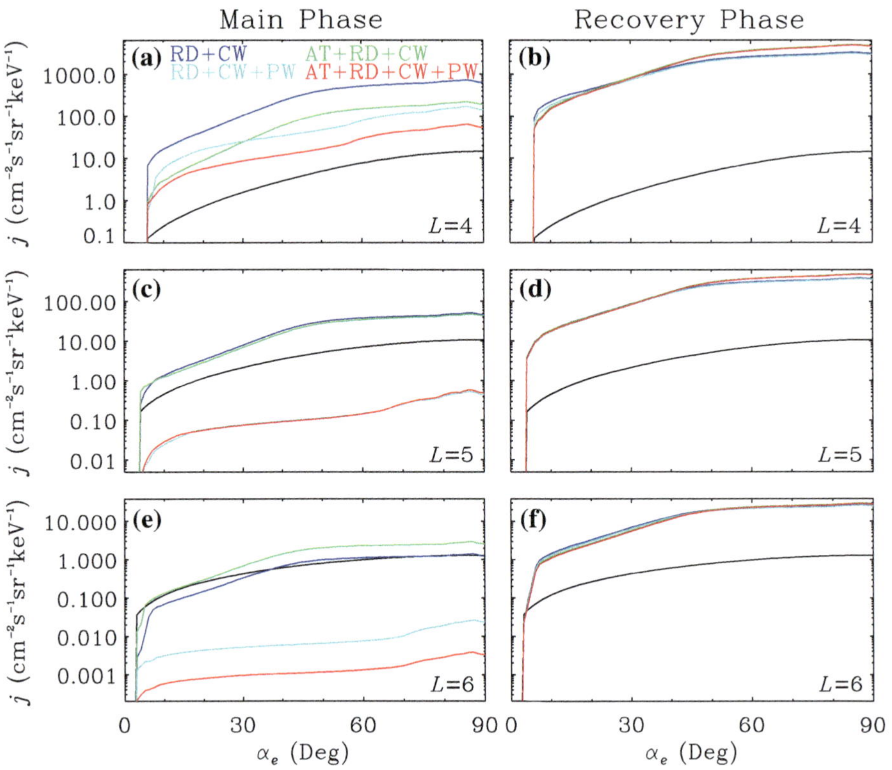

Fig. 4.11 Pitch-angle-dependent flux profiles for 1.0 MeV electrons at different L-shells (shown) during the main phase (*left*) and recovery phase (*right*). The *solid lines* represent the steady state. The *dotted*, *dashed*, *dot-dashed*, and *dot-dot-dashed lines* denote the simulation cases A1, A2, B1, and B2, respectively. In the panels (**b**), (**c**), (**d**), and (**f**), the *dotted* and *dot-dashed lines* overlap each other, and the *dashed* and *dot-dot-dashed lines* overlap each other

4.4 Application

4.4.1 Background

The outer radiation belt exhibits dramatic variabilities during geomagnetic storms (typically a rapid dropout during the main phase and a substantial buildup during the recovery phase). Over the past decades, significant progress has been made in the theory and simulation for the enhancements of radiation belt electron fluxes, but relatively limited progress has been reported on the depletions (see review by Friedel et al. 2002; Millan and Thorne 2007). In fact, the net effect of geomagnetic activity on radiation belt electron fluxes is ultimately determined by the competition between various source and loss processes (Reeves et al. 2003). Hence, both source and loss mechanisms must be understood fully in order to predict the radiation belt evolution

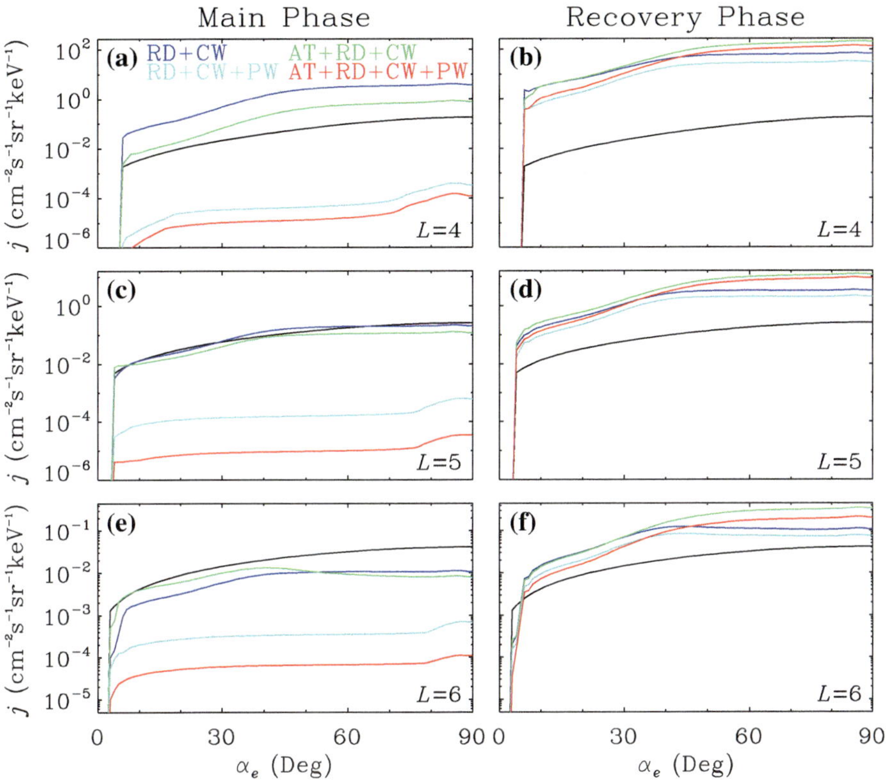

Fig. 4.12 The same as Fig. 4.11 except for 2.0 MeV electrons

(Green et al. 2004). Here we will concentrate on the radiation belt electron losses during storms.

There are four loss mechanisms invoked frequently in the early studies: Magnetopause shadowing (Desorgher et al. 2000), outward adiabatic transport (e.g., Kim and Chan 1997; Su et al. 2010b), outward radial diffusion (Shprits et al. 2006), and precipitation due to resonant interaction with a variety of plasma waves (chorus, hiss and EMIC) (e.g., Albert 2003; Summers and Thorne 2003; Glauert and Horne 2005; Thorne et al. 2005; Summers et al. 2007). A superposed epoch analysis (Green et al. 2004) suggested that enhanced precipitation to the atmosphere is the most promising mechanism for the depletion of relativistic electron fluxes, with an uncertainty on the exact cause of precipitation. A case study (Bortnik et al. 2006) indicated that the dominant loss mechanisms may be different at the inner and outer L-shells.

In this section, we study the 9 October 1990 electron radiation belt dropout event based on the observations and simulations. Note that some of the physical processes for outer radiation belt dynamics during the 9 October 1990 magnetic storm have been investigated in previous work (e.g., Brautigam and Albert 2000; Horne et al. 2003; Thorne et al. 2007; Albert et al. 2009).

4.4.2 Observations

An overview of the interplanetary and magnetospheric parameters (with data gaps) during the 9 October 1990 magnetic storm is shown in Fig. 4.13. The interplanetary southward magnetic field B_z (in GSM coordinates), solar wind flow pressure DP, and geomagnetic activity indices K_p and Dst are obtained from CDAweb-OMNI database (http://cdaweb.gsfc.nasa.gov/). The dayside magnetopause location L_{mp} is determined based on the B_z and DP parameters (Shue et al. 1998), and the plasmapause location L_{pp} is calculated based on the K_p index (Carpenter and Anderson 1992). The outbound passes of CRRES orbits 181–186 are also plotted. The CRRES satellite was launched on 25 July 1990 into a geosynchronous transfer orbit (an elliptical orbit with a perigee of 1.05 R_E, an apogee of 6.26 R_E, and an inclination of $\sim$18°). The orbital period of CRRES satellite was about 10 h. Our study primarily covers the time range from prestorm (orbit 181) to minimum-Dst (orbit 186). In this time period, there was an increase of the K_p index from 1 to 6, a decrease of the Dst index from 0 to -133 nT, a compression of the dayside magnetopause L_{mp} from 12 to 7 R_E, and a shrinking of the plasmapause L_{pp} from 5 to 3 R_E.

Figure 4.14 plots the radial profiles of radiation belt electron fluxes at the selected energy channels observed by the MEA instrument of CRRES during the orbits 181–186. Clearly, all of these electron flux profiles were relatively stable in the orbits 181–183. In orbit 184, the 0.148–1.178 MeV electron fluxes showed depletions of about a factor of 10–10^3 beyond 6 R_E. Considering the strong compression of the magnetopause (see Fig. 4.13) and relatively energy-independent characteristics of flux depletions, we think that such electron losses in the period between orbits 183 and 184 may result from the outward radial diffusion induced by the magnetopause shadowing effect. For orbit 185, there was a gradual loss of $>\sim$ 0.2 MeV electrons at the outer L-shells ($L > 5$), which can be explained by persistent outward radial diffusion and adiabatic transport. During orbit 186, 10–10^2 times depletions of electron fluxes at high energies (>0.4 MeV) were found to occur in the region $4.3 < L < 6.0$, likely related to the enhanced precipitation by EMIC and hiss waves in the high density plume (Bortnik et al. 2006).

4.4.3 Simulations

We next simulate this dropout event using the STEERB code. The main inputs of current STEERB code are the southward interplanetary magnetic field B_z, solar wind flow pressure DP, geomagnetic indices K_p and Dst, plume and chorus wave characteristics, initial condition, and outer boundary condition. The plume hiss and EMIC waves are assumed to be stimulated after CRRES orbit 185 with the longitude occurrence percentages 15 and 6 %, the latitude distribution range $\lambda \leq 40°$ and $\lambda \leq 15°$ and the radial distribution range $4.3 < L < 6.0$. The other wave parameters are listed in Table 3.1. The initial condition shown in Fig. 4.15 is the same as those adopted in the Sect. 4.3.1. The outer boundary condition is assumed to be

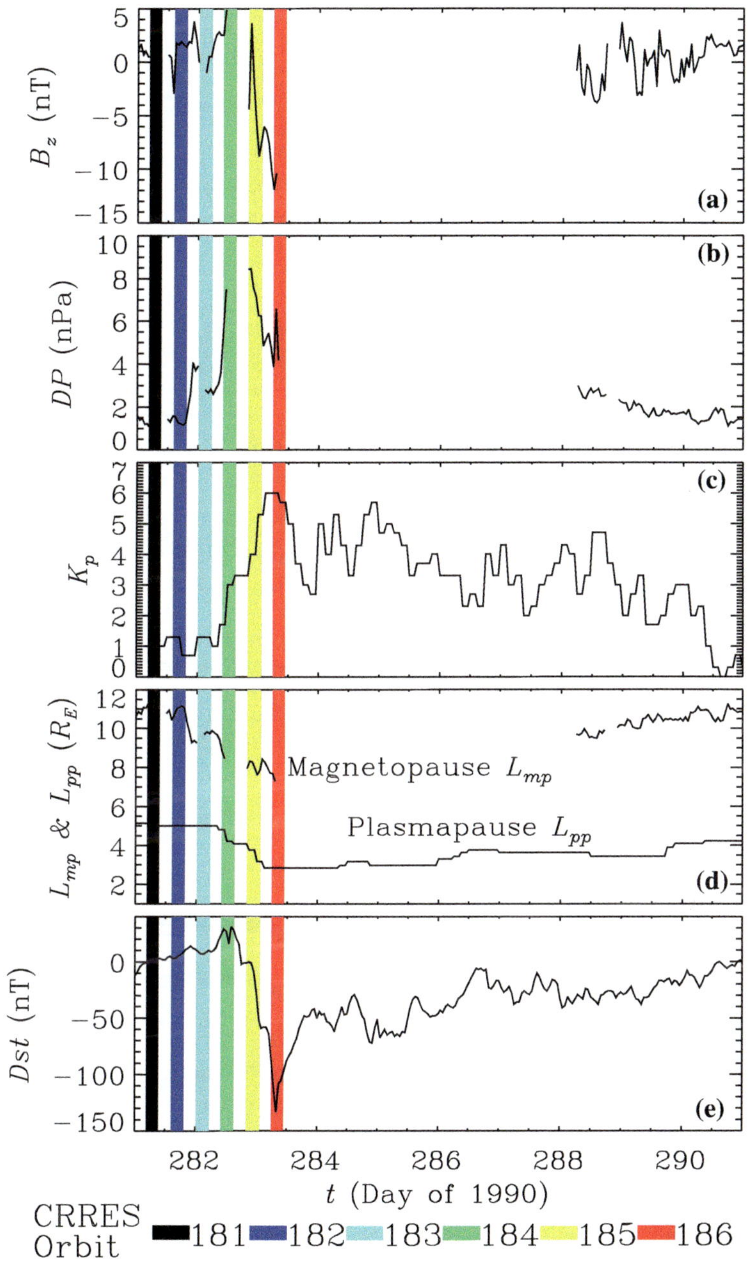

Fig. 4.13 **a** Southward interplanetary magnetic field in GSM coordinates B_z, **b** Solar wind flow pressure DP, **c** Geomagnetic activity index K_p, **d** Dayside magnetopause location L_{mp} computed based on B_z and DP (Shue et al. 1998), and plasmapause location L_{pp} computed based on K_p (Carpenter and Anderson 1992), **e** Geomagnetic activity index Dst. The outbound passes of CRRES orbits 181–186 are also plotted. Orbit 181 was in the prestorm period, and orbit 186 was in the minimum-Dst period

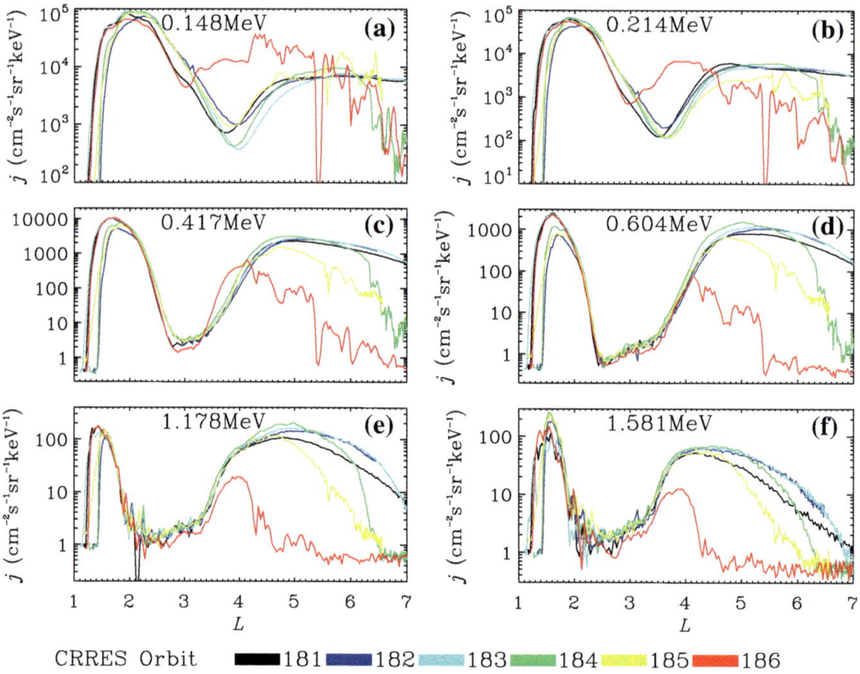

Fig. 4.14 Radial profiles of pitch-angle averaged electron fluxes at the selected energy channels observed by the MEA instrument of CRRES during orbits 181–186. Note that the actual latitude of the CRRES satellite changes with time

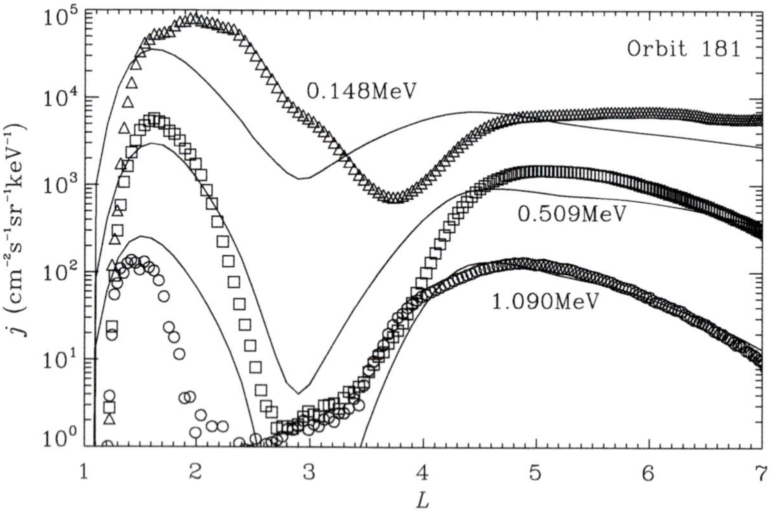

Fig. 4.15 STEERB simulations (*lines*) and CRRES observations (*symbols*) of pitch-angle averaged electron fluxes at the selected energy channels (shown) during the prestorm (orbit 181) period

$$f(t, \mu, J, L^* = 7) = f(t = 0, \mu, J, L^* = 7), \tag{4.20}$$

$$f(t = 0, \mu(\alpha_e, E_k), J(\alpha_e, E_k), L^* = 7) = F(E_k)\sin^n \alpha_e, \tag{4.21}$$

with $F(E_k)$ and q given in the Sect. 4.3.1.

Four simulation cases will be conducted to evaluate the contributions of different physical mechanisms: (A) MS + AT, (B) MS + AT + RD, (C) MS + AT + RD + PW, (D) MS + AT + RD + PW + CW, with MS, AT, RD, PW and CW referring to the magnetopause shadowing, adiabatic transport, radial diffusion, and plume and chorus wave-particle interactions, respectively. Note that the magnetopause shadowing process is included by removing the energetic (>0.1 MeV) electrons in the region $L > L_{mp}$. Figures 4.15 and 4.16 shows the radial flux profiles of STEERB simulations for these cases at the energy channels 0.148, 0.509, and 1.090 MeV, and the corresponding CRRES observations are also plotted for comparison. It is clear that the STEERB simulations are in good agreement with CRRES observations of energetic outer radiation belt electron fluxes during the prestorm period (orbit 181). The model to data agreement is relatively poor in the slot region and inner belt, which may not significantly affect the subsequent model evolution of outer radiation belt electrons. In the minimum-Dst period (orbit 186), the goodness of fit between observations and simulations for the first three test cases is found to increase with the number increasing of physical mechanisms included. Combined magnetopause shadowing and adiabatic transport tend to underestimate these energetic electron fluxes in the inner L-shells ($L < 5$), and overestimate them in the outer L-shells ($L > 5$). Incorporation of radial diffusion can well reproduce the low energy (0.148 MeV) electron flux, but overestimate the high-energy electron fluxes throughout the outer radiation belt. The further introduction of plume wave-particle interactions is able to make the simulations agree well with the observations at all the energy channels especially outside of $L = 4$. The notable disagreement for 0.509 MeV electron flux around $L = 3.5$ (where no obvious depletion occurred) primarily results from the errors of initial condition. Note that the plume wave-particle interactions mainly affect the high-energy electron fluxes. Comparing with Case C, the last addition of chorus waves does not significantly change the electron fluxes. These numerical results indicate that the current electron radiation belt dropout event can be generally explained by the competition and combination of all these physical mechanisms, with the dominance of the first four physical processes (MS, AT, RD, and PW).

4.5 Conclusions and Discussions

We introduce the fully adiabatic transport process into the STEERB code (Su et al. 2010a) by adopting the Dst-dependent Hilmer-Voigt symmetric geomagnetic field model (Hilmer and Voigt 1995). The current STEERB model includes more complete physical processes: adiabatic transport, radial diffusion, and various in-situ wave-particle interactions. The present STEERB code are implemented based on the same

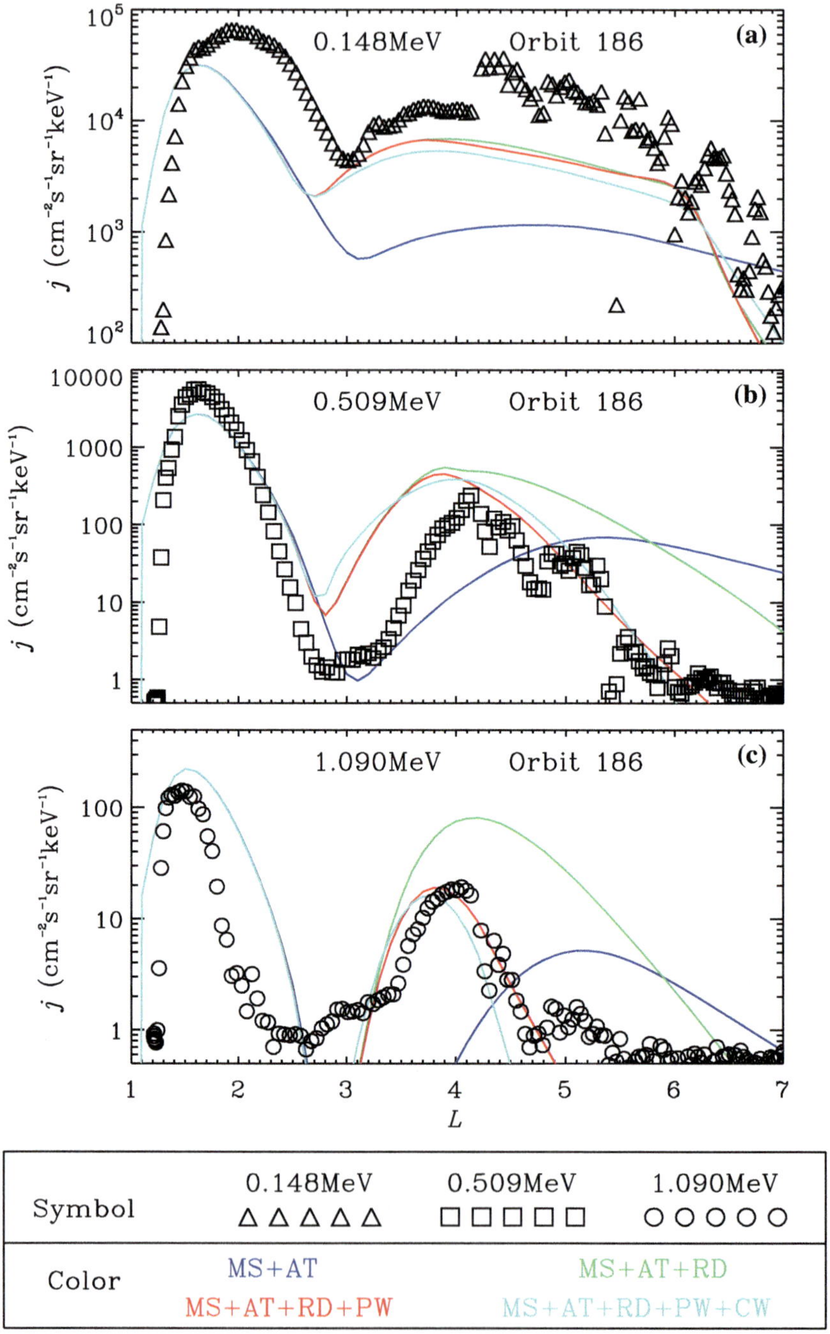

Fig. 4.16 STEERB simulations (*lines*) and CRRES observations (*symbols*) of pitch-angle averaged electron fluxes at the selected energy channels (shown) during the minimum-*Dst* (orbit 186) period

numerical scheme as that used in the previous model (see Chap. 3), and still keeps efficient, stable, and easy to parallelize.

We further evaluate the influence of adiabatic transport on radiation belt electron dynamics based some idealized simulations:

1. The adiabatic transport is a fully reversible process. The outer radiation belt electron fluxes decrease by a factor of above 60 during the main phase, and gradually return to the prestorm state during the recovery phase. With the decrease in Dst index, the electron pitch-angle distribution can evolve from the 90°-peaked distribution into the butterfly-shaped distribution.
2. In the presence of strong chorus-driven acceleration process, the adiabatic transport alone (without plume waves) is found to be hard-pressed to produce the expected main phase loss of energetic outer radiation belt electron fluxes. Note that the adiabatic transport is able to cause significant depletion of energetic electrons beyond $5R_E$ in the absence of chorus waves (Su et al. 2010b).
3. These adiabatic and nonadiabatic processes for radiation belt electron dynamics are coupled to each other. The adiabatic transport, together with radial diffusion and cyclotron resonant interactions with chorus and plume waves contribute significantly to the main phase loss and the recovery phase enhancement of energetic electron fluxes. The ignoring of adiabatic transport generally produces the overestimate (underestimate) of energetic outer radiation belt electron fluxes during the main (recovery) phase of a storm. In the current simulations, the main phase overestimate factor is up to 5–30 over all the pitch-angles, and the recovery phase underestimate factor can approach 2–5 at larger pitch-angels ($\alpha_e > 50°$). These numerical results suggest that the adiabatic transport should be incorporated in future radiation belt models for space weather application.

The electron radiation belt dropout event during the 9 October 1990 magnetic storm is studied based on the observations and simulations.

1. The CRRES satellite provides a good representation of the radiation belt electron signatures from the prestorm (orbit 181) to minimum-Dst periods (orbit 186). This dropout event is found to display different characteristics at different L-shells: In the region $6.0 < L < 7.0$, significant (10–10^3 times) depletions occurred at ~0.1–1.0 MeV electron fluxes; in the region $4.3 < L < 6.0$, large (10–10^2 times) depletions were restricted to the high-energy (>0.4 MeV) electron fluxes. Such dropout characteristics are similar to those of the 20 November 2003 event (Bortnik et al. 2006).
2. The three-dimensional kinetic radiation belt model STEERB is adopted to simulate this specific dropout event, with the inputs of interplanetary parameters, geomagnetic indices, plume and chorus wave characteristics, initial condition, and outer boundary condition. The STEERB code, with incorporation of the magnetopause shadowing, adiabatic transport, radial diffusion, and plume and chorus wave-particle interactions, can basically reproduce the observed ~0.1–1.0 MeV electron flux depletion in the outer radiation belt, suggesting that the competition and combination of all these physical mechanisms can well explain the loss of radiation belt electrons.

References

Albert JM (2003) Evaluation of quasi-linear diffusion coefficients for EMIC waves in a multispecies plasma. J Geophys Res 108(A8):1249. doi:10.1029/2002JA009792

Albert JM, Meredith NP, Horne RB (2009) Three-dimensional diffusion simulation of outer radiation belt electrons during the october 9, 1990, magnetic storm. J Geophys Res 114(A09):214. doi:10.1029/2009JA014336

Bortnik J, Thorne RM, O'Brien TP, Green JC, Strangeway RJ, Shprits YY, Baker DN (2006) Observation of two distinct, rapid loss mechanisms during the 20 November 2003 radiation belt dropout event. J Geophys Res 111(A12):216. doi:10.1029/2006JA011802

Brautigam DH, Albert JM (2000) Radial diffusion analysis of outer radiation belt electrons during the October 9, 1990, magnetic storm. J Geophys Res 105:291–310. doi:10.1029/1999JA900344

Carpenter DL, Anderson RR (1992) An ISEE/whistler model of equatorial electron density in the magnetosphere. J Geophys Res 97:1097–1108

Desorgher L, Bühler P, Zehnder A, Flückiger EO (2000) Simulation of the outer radiation belt electron flux decrease during the March 26, 1995, magnetic storm. J Geophys Res 105:21211–21224. doi:10.1029/2000JA900060

Dessler AJ, Karplus R (1961) Some effects of diamagnetic ring currents on Van Allen radiation. J Geophys Res 66:2289–2295. doi:10.1029/JZ066i008p02289

Friedel RHW, Reeves GD, Obara T (2002) Relativistic electron dynamics in the inner magnetosphere—a review. J Atmos Sol-Terr Phys 64:265–282. doi:10.1016/S1364-6826(01)00088-8

Ganushkina NY, Pulkkinen TI, Kubyshkina MV, Singer HJ, Russell CT (2002) Modeling the ring current magnetic field during storms. J Geophys Res 107:1092. doi:10.1029/2001JA900101

Glauert SA, Horne RB (2005) Calculation of pitch angle and energy diffusion coefficients with the PADIE code. J Geophys Res 110(A04):206. doi:10.1029/2004JA010851

Green JC, Onsager TG, O'Brien TP, Baker DN (2004) Testing loss mechanisms capable of rapidly depleting relativistic electron flux in the Earth's outer radiation belt. J Geophys Res 109(A12):211. doi:10.1029/2004JA010579

Hilmer RV, Voigt G (1995) A magnetospheric magnetic field model with flexible current systems driven by independent physical parameters. J Geophys Res 100:5613–5626. doi:10.1029/94JA03139

Horne RB, Meredith NP, Thorne RM, Heynderickx D, Iles RHA, Anderson RR (2003) Evolution of energetic electron pitch angle distributions during storm time electron acceleration to megaelectronvolt energies. J Geophys Res 108:1016. doi:10.1029/2001JA009165

Kim H-J, Chan AA (1997) Fully adiabatic changes in storm time relativistic electron fluxes. J Geophys Res 102:22107–22116. doi:10.1029/97JA01814

Kim KC, Lee D, Kim H, Lee ES, Choi CR (2010) Numerical estimates of drift loss and dst effect for outer radiation belt relativistic electrons with arbitrary pitch angle. J Geophys Res 115(A03):208. doi:10.1029/2009JA014523

Li W, Shprits YY, Thorne RM (2007) Dynamic evolution of energetic outer zone electrons due to wave-particle interactions during storms. J Geophys Res 112(A11):A10, 220. doi: 10.1029/2007JA012368

McIlwain CE (1966) Ring current effects on trapped particles. J Geophys Res 71:3623–3628

Millan RM, Thorne RM (2007) Review of radiation belt relativistic electron losses. J Atmos Sol-Terr Phys 69:362–377. doi:10.1016/j.jastp.2006.06.019

Reeves GD, McAdams KL, Friedel RHW, O'Brien TP (2003) Acceleration and loss of relativistic electrons during geomagnetic storms. Geophys Res Lett 30:1529. doi:10.1029/2002GL016513

Schulz M, Lanzerotti LJ (1974) Particle diffusion in the radiation belts. Physics and chemistry in space, vol 7. Springer, New York

Shprits YY, Thorne RM, Friedel R, Reeves GD, Fennell J, Baker DN, Kanekal SG (2006) Outward radial diffusion driven by losses at magnetopause. J Geophys Res 111(A11):214. doi:10.1029/2006JA011657

Shprits YY, Subbotin D, Ni B (2009) Evolution of electron fluxes in the outer radiation belt computed with the VERB code. J Geophys Res 114(A11):209. doi:10.1029/2008JA013784
Shue J, Song P, Russell CT, Steinberg JT, Chao JK, Zastenker G, Vaisberg OL, Kokubun S, Singer HJ, Detman TR, Kawano H (1998) Magnetopause location under extreme solar wind conditions. J Geophys Res 103:17691–17700. doi:10.1029/98JA01103
Su Z, Zheng H, Wang S (2009a) Dynamic evolution of energetic outer zone electrons due to whistler-mode chorus based on a realistic density model. J Geophys Res 114(A07):201. doi:10.1029/2008JA014013
Su Z, Zheng H, Wang S (2009b) Evolution of electron pitch angle distribution due to interactions with whistler-mode chorus following substorm injections. J Geophys Res 114(A08):202. doi:10.1029/2009JA014269
Su Z, Zheng H, Xiong M (2009c) Dynamic evolution of outer radiation belt electrons due to whistler-mode chorus. Chin Phys Lett 26(039):401
Su Z, Xiao F, Zheng H, Wang S (2010a) STEERB: a three-dimensional code for storm-time evolution of electron radiation belt. J Geophys Res 115(A09):208. doi:10.1029/2009JA015210
Su Z, Xiao F, Zheng H, Wang S (2010b) Combined radial diffusion and adiabatic transport of radiation belt electrons with arbitrary pitch-angles. J Geophys Res 115(A10):249. doi:10.1029/2010JA015903
Su Z, Zheng H, Wang S (2010c) A parametric study on the diffuse auroral precipitation by resonant interaction with whistler-mode chorus. J Geophys Res 115(A05):219. doi:10.1029/2009JA014759
Su Z, Zheng H, Wang S (2010d) Three dimensional simulations of energetic outer zone electron dynamics due to wave-particle interaction and azimuthal advection. J Geophys Res 115(A06):203. doi:10.1029/2009JA014980
Su Z, Xiao F, Zheng H, Wang S (2011a) CRRES observation and STEERB simulation of the 9 (October 1990) electron radiation belt dropout event. Geophys Res Lett 38(L06):106. doi:10.1029/2011GL046873
Su Z, Xiao F, Zheng H, Wang S (2011b) Radiation belt electron dynamics driven by adiabatic transport, radial diffusion, and wave-particle interactions. J Geophys Res 116(A04):205. doi:10.1029/2010JA016228
Su Z, Zheng H, Chen L, Wang S (2011c) Numerical simulations of storm-time outer radiation belt dynamics by wave-particle interactions including cross diffusion. J Atmos Sol-Terr Phys 73:95–105. doi:10.1016/j.jastp.2009.08.002
Summers D, Thorne RM (2003) Relativistic electron pitch-angle scattering by electromagnetic ion cyclotron waves during geomagnetic storms. J Geophys Res 108:1143. doi:10.1029/2002JA009489
Summers D, Ni B, Meredith NP (2007) Timescales for radiation belt electron acceleration and loss due to resonant wave-particle interactions: 1. Theory J Geophys Res 112(A04):206. doi:10.1029/2006JA011801
Thorne RM, O'Brien TP, Shprits YY, Summers D, Horne RB (2005) Timescale for MeV electron microburst loss during geomagnetic storms. J Geophys Res 110(A09):202. doi:10.1029/2004JA010882
Thorne RM, Shprits YY, Meredith NP, Horne RB, Li W, Lyons LR (2007) Refilling of the slot region between the inner and outer electron radiation belts during geomagnetic storms. J Geophys Res 112(A06):203. doi:10.1029/2006JA012176
Tsyganenko NA (1995) Modeling the earth's magnetospheric magnetic field confined within a realistic magnetopause. J Geophys Res 100:5599–5612
Xiao F, Su Z, Zheng H, Wang S (2009) Modeling of outer radiation belt electrons by multidimensional diffusion process. J Geophys Res 114(A03):201. doi:10.1029/2008JA013580
Xiao F, Su Z, Zheng H, Wang S (2010) Three-dimensional simulations of outer radiation belt electron dynamics including cross diffusion terms. J Geophys Res 115(A05):216. doi:10.1029/2009JA014541

Chapter 5
Magnetospheric Convection

5.1 Introduction

The trapped particles in the inner magnetosphere have three adiabatic invariants associated to the three periodic motions (see Sect. 1.3.1). Consequently, the physical processes controlling the radiation belt evolution can be classified into adiabatic and non-adiabatic processes. The adiabatic transport process has been investigated in Chap. 4, and two non-adiabatic processes, cyclotron, and drift resonances have been examined in Chaps. 2 and 3. In fact, there remains another non-adiabatic process, magnetospheric convection, on a timescale smaller than the drift-period but larger than the bounce and cyclotron periods.

Bourdarie et al. (1997) constructed the first radiation belt convection-diffusion model, including the magnetospheric convection and the pitch-angle diffusion by cyclotron resonance and Coulomb collision. Recently, some new convection-diffusion models for electron radiation belt were generalized from the ring current models (Fok et al. 1995, 2010, 2011; Fok and Moore 1997; Jordanova et al. 1996, 1997, 2003, 2008, 2010), have been reported, e.g., RAM (Ring current-Atmosphere interactions Model) (Jordanova and Miyoshi 2005; Miyoshi et al. 2006) and RBE (Radiation Belt Environment) (Fok et al. 2008).

In this chapter, the magnetospheric convection is introduced to our previously developed global electron radiation belt diffusion model, yielding the global radiation belt convection-diffusion model STEERB. The data-driven STEERB model was further used to simulate the substorm injection on 10 January 1997 and quantify its contribution to radiation belt evolution.

Z. Su, *A Global Kinetic Model for Electron Radiation Belt Formation and Evolution*, Springer Theses, DOI 10.1007/978-3-662-46651-3_5

5.2 Global Radiation Belt Convection-Diffusion Model STEERB

5.2.1 Background Electromagnetic Fields

The global geomagnetic field is characterized by the *Dst*-dependent Hilmer-Voigt geomagnetic field model (Hilmer and Voigt 1995). It should be mentioned that, the diffusion coefficients for cyclotron resonance are still calculated in a fixed dipolar magnetic field to save computing time.

The equatorial electric field $\mathbf{E_e}$ consists of the convection electric field $\mathbf{E_{conv}}$, the co-rotation electric field $\mathbf{E_{cor}}$, and the impulsive electric field $\mathbf{E_i}$. The convection electric field is described by the Volland-Maynard-Chen model (Volland 1973; Maynard and Chen 1975)

$$\mathbf{E_{conv}} = -\nabla\Phi, \tag{5.1}$$

$$\Phi = AL^2 \sin\phi, \tag{5.2}$$

$$A = \frac{45}{(1 - 0.159K_p + 0.0093K_p^2)^3}\ [\text{V}], \tag{5.3}$$

where ϕ is the longitude with $\phi = 0$ for midnight and $\phi = 90°$ for dawn. In a spherical coordinate system $(r,\ \theta,\ \phi)$, the co-rotation electric field can be written as

$$\mathbf{E_{cor}} = -\left(\boldsymbol{\omega}_\mathbf{E} \times \mathbf{r}\right) \times \mathbf{B_e}, \tag{5.4}$$

where $\boldsymbol{\omega}_\mathbf{E}$ denote the terrestrial rotation angular velocity, and $\mathbf{B_e}$ equatorial magnetic field. The substorm-associated impulsive electric field is depicted as an Earthward propagating pulse model (Sarris et al. 2002)

$$\mathbf{E_i} = -\mathbf{e}_\phi E_0 \left[1 + c_1 \cos(\phi - \phi_0)\right]^{n_\phi} \exp\left(-\xi^2\right), \tag{5.5}$$

$$\xi = \left[r - r_i + v(r)(t - t_a)\right]/d, \tag{5.6}$$

$$v(r) = a + br, \tag{5.7}$$

$$t_a = c_2 R_E / v_a \left[1 - \cos(\phi - \phi_0)\right], \tag{5.8}$$

where E_0 controls the strength of pulse; ξ determines the maxima location; d is the pulse width; $v(r)$ is the pulse front speed as a function of r; c_1 and n_ϕ characterize the dependence of electric amplitude on ϕ; t_a represents the delay of the pulse from ϕ_0 to other longitudes; c_2 affects the magnitude of the delay; v_a is the longitudinal propagation speed of the pulse.

5.2.2 Basic Equation

In a symmetrical geomagnetic field model, the bounce-averaged convection-diffusion kinetic equation for radiation belt electron PSD f evolution can be written as (e.g., Schulz and Lanzerotti 1974; Bourdarie et al. 1997; Fok et al. 2001; Su et al. 2010b, 2011a, b)

$$\begin{aligned}
&\frac{\partial f}{\partial t}+\left\langle\frac{\mathrm{d}L^*}{\mathrm{d}t}\right\rangle\left.\frac{\partial f}{\partial L^*}\right|_{\mu,J,\phi}+\left\langle\frac{\mathrm{d}\phi}{\mathrm{d}t}\right\rangle\left.\frac{\partial f}{\partial \phi}\right|_{\mu,J,L^*}\\
&=\frac{1}{Gp}\left.\frac{\partial}{\partial\alpha_e}\right|_{p,L,\phi}\left[G\left(\langle D_{\alpha\alpha}\rangle\frac{1}{p}\left.\frac{\partial f}{\partial\alpha_e}\right|_{p,L,\phi}+\langle D_{\alpha p}\rangle\left.\frac{\partial f}{\partial p}\right|_{\alpha_e,L,\phi}\right)\right]\\
&+\frac{1}{G}\left.\frac{\partial}{\partial p}\right|_{\alpha_e,L,\phi}\left[G\left(\langle D_{p\alpha}\rangle\frac{1}{p}\left.\frac{\partial f}{\partial\alpha_e}\right|_{p,L,\phi}+\langle D_{pp}\rangle\left.\frac{\partial f}{\partial p}\right|_{\alpha_e,L,\phi}\right)\right]\\
&+L^{*2}\left.\frac{\partial}{\partial L^*}\right|_{\mu,J,\phi}\left(\frac{D_{L^*L^*}}{L^{*2}}\left.\frac{\partial f}{\partial L^*}\right|_{\mu,J,\phi}\right)-\frac{f}{\tau_L}-\frac{f}{\tau_C}-\frac{f}{\tau_W}.
\end{aligned} \tag{5.9}$$

The STEERB model covers two computational spaces: action-angle phase space (μ, J, L^*, ϕ) and observable space (α_e, p, L, ϕ). Similar to that in Chap. 4, the mapping from action-angle phase space to observable space can self-consistently incorporate the adiabatic transport process due to magnetic field variation. The left-hand side of Eq. (5.9) describes the change in electron PSD along the drift paths. The right-hand side of Eq. (5.9) treats the local diffusion by cyclotron resonance (see Chap. 2), the radial diffusion by drift-resonance (see Chap. 3), and the empirical loss processes (see Chap. 3) in the loss cone, near the Earth and in the slot region.

5.2.3 Convection and Diffusion Coefficients

The bounce-averaged drift velocity (Roederer 1970) in vector form can be expressed as

$$\begin{aligned}
\langle\mathbf{V_e}\rangle&=\left\langle\frac{\mathrm{d}L}{\mathrm{d}t}\right\rangle R_E\mathbf{e_r}+\left\langle\frac{\mathrm{d}\phi}{\mathrm{d}t}\right\rangle LR_E\mathbf{e}_\phi\\
&=\frac{\mathrm{d}L}{\mathrm{d}L^*}\left\langle\frac{\mathrm{d}L^*}{\mathrm{d}t}\right\rangle R_E\mathbf{e_r}+\left\langle\frac{\mathrm{d}\phi}{\mathrm{d}t}\right\rangle LR_E\mathbf{e}_\phi\\
&=\frac{\mathbf{E_e}\times\mathbf{B_e}}{B_e^2}-\frac{2p}{e\tau_B B_e^2}\nabla_e I\times\mathbf{B_e}+\omega_E LR_E\mathbf{e}_\phi,
\end{aligned} \tag{5.10}$$

representing the electric, gradient, and curvature drifts (see Sect. 1.3.1). The diffusion coefficients in the equatorial pitch-angle, momentum, cross term, and Roederer L^* are exactly the same as those adopted in Chap. 4

5.2.4 Numerical Method

Based on a split-operator technique, the solution of four-dimensional convection–diffusion Eq. (5.9) is advanced in three steps. The cyclotron resonant diffusion processes

$$\frac{\partial f}{\partial t}=\frac{1}{Gp}\frac{\partial}{\partial \alpha_e}\bigg|_{p,L,\phi}\left[G\left(\langle D_{\alpha\alpha}\rangle\frac{1}{p}\frac{\partial f}{\partial \alpha_e}\bigg|_{p,L,\phi}+\langle D_{\alpha p}\rangle\frac{\partial f}{\partial p}\bigg|_{\alpha_e,L,\phi}\right)\right] \tag{5.11}$$
$$+\frac{1}{G}\frac{\partial}{\partial p}\bigg|_{\alpha_e,L,\phi}\left[G\left(\langle D_{p\alpha}\rangle\frac{1}{p}\frac{\partial f}{\partial \alpha_e}\bigg|_{p,L,\phi}+\langle D_{pp}\rangle\frac{\partial f}{\partial p}\bigg|_{\alpha_e,L,\phi}\right)\right]-\frac{f}{\tau_L},$$

are solved by the HFD method with the same boundary conditions as those in Chap. 4. The radial convection and diffusion processes with empirical loss terms

$$\frac{\partial f}{\partial t}+\left\langle\frac{\mathrm{d}L^*}{\mathrm{d}t}\right\rangle\frac{\partial f}{\partial L^*}\bigg|_{\mu,J,\phi}=L^{*2}\frac{\partial}{\partial L^*}\bigg|_{\mu,J,\phi}\left(\frac{D_{L^*L^*}}{L^{*2}}\frac{\partial f}{\partial L^*}\bigg|_{\mu,J,\phi}\right)-\frac{f}{\tau_C}-\frac{f}{\tau_W}, \tag{5.12}$$

are solved by the traditional fully-implicit scheme with the boundary conditions

$$f(L^*=1)=0, \tag{5.13}$$

$$\begin{cases} f(L^*=7)=f_o(t,\mu,J,\phi), & \text{when}\left\langle\frac{\mathrm{d}L^*}{\mathrm{d}t}\right\rangle\le 0 \\ \frac{\partial f}{\partial L^*}(L^*=7)=0, & \text{when}\left\langle\frac{\mathrm{d}L^*}{\mathrm{d}t}\right\rangle> 0 \end{cases}. \tag{5.14}$$

The azimuthal convection process

$$\frac{\partial f}{\partial t}+\left\langle\frac{\mathrm{d}\phi}{\mathrm{d}t}\right\rangle\frac{\partial f}{\partial \phi}\bigg|_{\mu,J,L^*}=0 \tag{5.15}$$

is solved by the Weighted Essentially Non-Oscillatory (WENO) scheme (Shu and Osher 1989) with the periodic boundary condition

$$f(\phi=0^\circ)=f(\phi=360^\circ). \tag{5.16}$$

Note that the WENO scheme, a shock-capturing method, is vital to capture the transient phenomenons without notable numerical oscillations.

5.3 Application

5.3.1 Background

The substorm injection process can produce rapid enhancement of radiation belt electron fluxes on a timescale of minutes to hours (McIlwain et al. 1974; Friedel et al. 1996; Reeves et al. 1996). Successful test-particle simulations of electron substorm injections up to several hundred keV have been presented in a dynamic three-dimensional MHD electromagnetic field (Birn et al. 1998) or in a static magnetic field (without background electric fields) superimposed by an impulsive earthward propagating electromagnetic field (e.g., Li et al. 1998; Sarris et al. 2002; Zaharia et al. 2004). Until now, there are relatively few kinetic radiation belt models quantitatively handling the substorm injection processes. Fok et al. (2001) only performed some idealized kinetic simulations of substorm injection without any quantitative comparison to satellite observations. In this section, we use the convection–diffusion STEERB to simulate the radiation belt electron evolution driven by substorm injection on 10 January 1997. In view of the relatively short timescale (several hours), the cyclotron resonance is ignored in the present simulations.

5.3.2 Inputs

The initial condition is assumed to be isotropic in local time, which is obtained by solving the steady radial diffusion equation (Su et al. 2010a, 2011b). Figure 5.1a shows the global distribution of radiation belt electron fluxes, and Fig. 5.1b plots the energy spectrum at $L = 6.6$, consistent with the observations of SOPA instrument onboard LANL 1990-095.

The nightside outer boundary condition is set to be coherent with the substorm-related impulsive electric field E_i at $L^* = 7.0$:

$$f_o(t, \mu, J, \phi) = \bar{f}_o(t = 0, \mu, J) \cdot f_r(t, \mu, J, \phi), \tag{5.17}$$

$$f_r(t, \mu, J, \phi) = 1 + \left(\bar{f}_r(\mu, J) - 1\right) \cdot \frac{|E_i(L^* = 7, \phi)|}{\max(|E_i(L^*, \phi)|)}, \tag{5.18}$$

$$\bar{f}_o(t = 0, \mu, J) = c_\kappa \left[1 + \frac{E_k}{\kappa E_0}\right]^{-(\kappa+1)} \sin^q \alpha_e, \tag{5.19}$$

$$\bar{f}_r(\mu, J) = \frac{j_2(E_k(\mu, J))}{j_1(E_k(\mu, J))} \tag{5.20}$$

where $\bar{f}_o(t = 0, \mu, J)$ is specified by a kappa distribution with $\kappa = 3.0$, $E_0 = 3$ keV and $q = 0.855$ (Thorne et al. 2005); $\bar{f}_r(\mu, J)$ is determined with the similar technique of Brautigam and Albert (2000); j_1 and j_2 represent the quiet and maximal injected electron fluxes observed by SOPA instrument onboard LANL 1990-095 satellite.

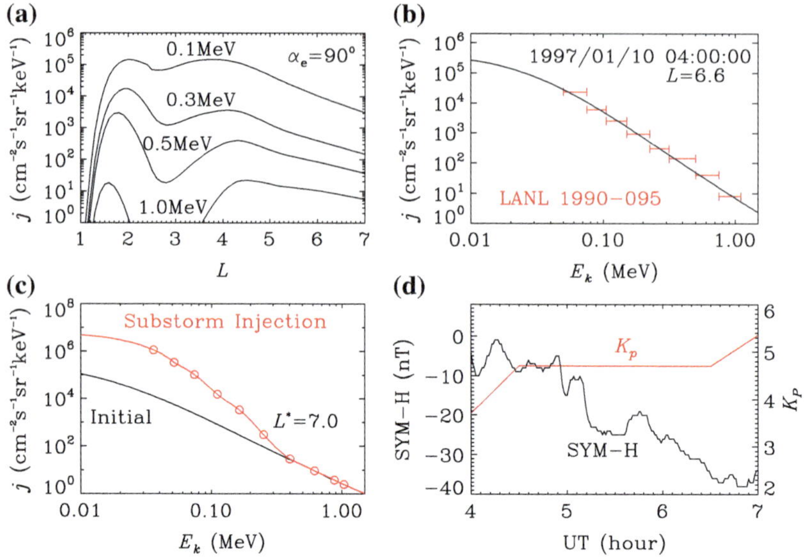

Fig. 5.1 **a** Initial radial profiles of equatorially-trapped electron flux at different kinetic energies (shown); **b** initial energy spectrum modeled by STEERB (*solid line*) and observed by SOPA of LANL 1990-095 (*red symbol*) at $L = 6.6$; **c** energy spectrum at the outer boundary $L^* = 7$ during the quiet period (*black*) and the most significant injection period (*red*); **d** SYM-H (*black*) and K_p (*red*) indices during 0400-0700 UT on 10 January 1997

Figure 5.1c shows the energy spectrum at the outer boundary during the quiet period and the most significant injection period.

Figure 5.1d plots the SYM-H and K_p indices during the simulation time range. The SYM-H index, possessing high time resolution, is substituted for Dst index to drive the Hilmer-Voigt symmetric magnetic field model in the present study. The K_p index is used to determine the background convection electric field and radial diffusion coefficients.

The free parameters for impulsive electric field are set to be: $c_1 = 1, c_2 = 0.5, a = 53.15$ km/s, $b = 0.0093\,\text{s}^{-1}$, $n = 3$, $v_a = 20$ km/s, $r_i = 150.2R_E$, $d = 8 \times 10^7$ m, $E_0 = 2.5/2^n$ mV/m and $\phi_0 = 0$. After numerous test simulations, these parameters appear to give the optimal results.

5.3.3 Outputs

Figure 5.2 shows the global distributions and the azimuthally averaged radial profiles of 0.1 and 0.5 MeV electron fluxes at the selected times. The freshly injected energetic electrons at nightside convect earthward (primarily due to the impulsive electric field), simultaneously drift eastward to dayside and disperse in local times (faster for higher energy electrons). At 0500UT, the azimuthally averaged energetic electron

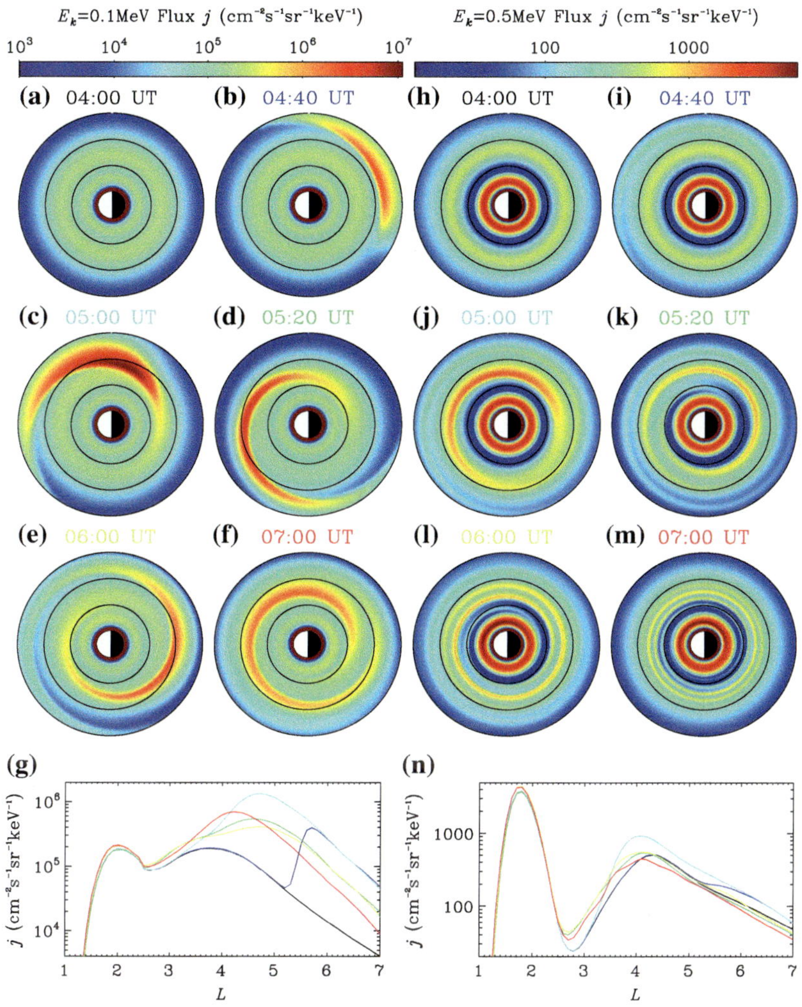

Fig. 5.2 Evolution of equatorially trapped 0.1 MeV (**a–f**) and 0.5 MeV (**h–m**) electron flux in the (L, ϕ) space, and the corresponding azimuthally averaged radial flux profiles at 0.1 MeV (**g**) and 0.5 MeV (**n**). The *black circles* denote the locations $L = 1$, 3, 5 and 7. The magnetic local noon is on the *left*, dusk is at the *bottom*, midnight is on the *right*, and dawn is at the *top*

fluxes show the most significant enhancement ($\sim$10 times at 0.1 MeV and $\sim$2 times at 0.5 MeV) at the center of outer radiation belt. At the moment, the impulsive electric field has traveled inside $L = 3.5$, and will contribute little to the subsequent evolution of outer radiation belt. After 0500UT, there is a gradual decrease of energetic outer zone electron fluxes resulted from the outward adiabatic transport (see the SYM-H index in Fig. 5.1d) (Su et al. 2010b) and the enhanced convective loss through

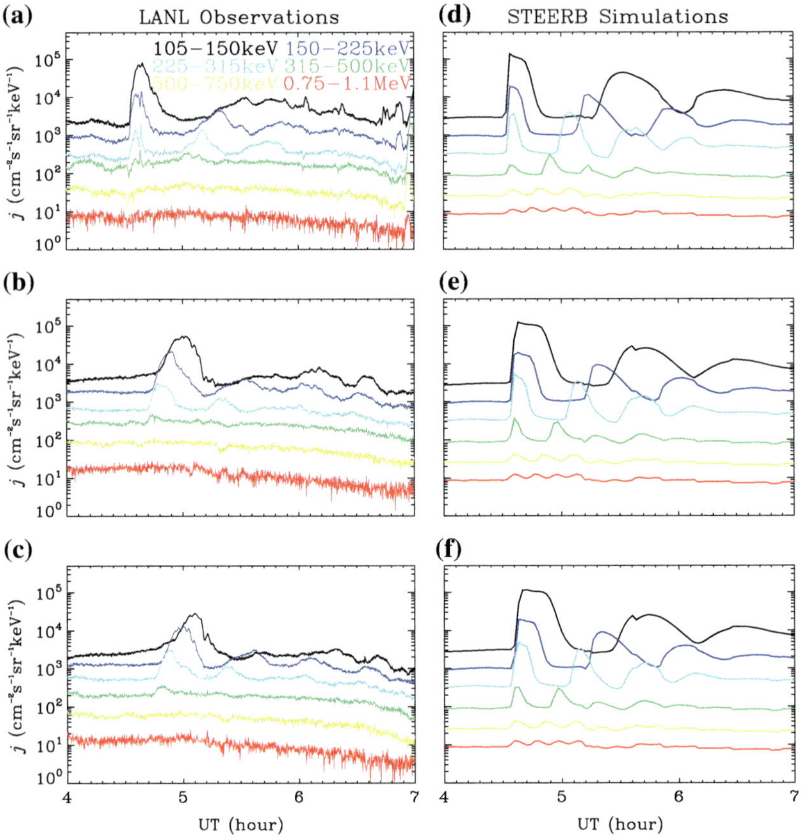

Fig. 5.3 Evolution of energetic electron fluxes measured by three different LANL satellites (*left*) and computed by STEERB model (*right*) at each satellite position. **a** and **d** correspond to LANL 1990-095 (LT = UT − 0230), **b** and **e** correspond to LANL 1991-080 (LT = UT + 0442), **c** and **f** correspond to LANL 1994-084 (LT = UT + 0654)

dayside boundary (see the K_p index in Fig. 5.1d). In fact, the radial diffusion process is relatively unimportant on such a short injection timescale (∼hours).

Figure 5.3 shows the comparison between observed and modeled energetic electron fluxes at three different LANL satellite positions. There are notable injections of <500 keV electrons at the geostationary orbit. Some injected energetic electrons can continually reappear at a spacecraft location, which is called "drift echo". The fresh injection is relatively "dispersionless" (i.e., the rise of electron fluxes can be observed almost simultaneously over a wide energy range), and subsequently becomes dispersive owing to the differential drift. The STEERB simulations can basically reproduce the enhancement level, drift echo, and dispersion feature observed by LANL satellites especially at nightside (LANL 1990-095). As a result of the geomagnetic field symmetry, the simulated injection fronts appear to arrive earlier than the observed ones at dayside (LANL 1991-080 and 1994-084). In a realistic asymmetric geomagnetic

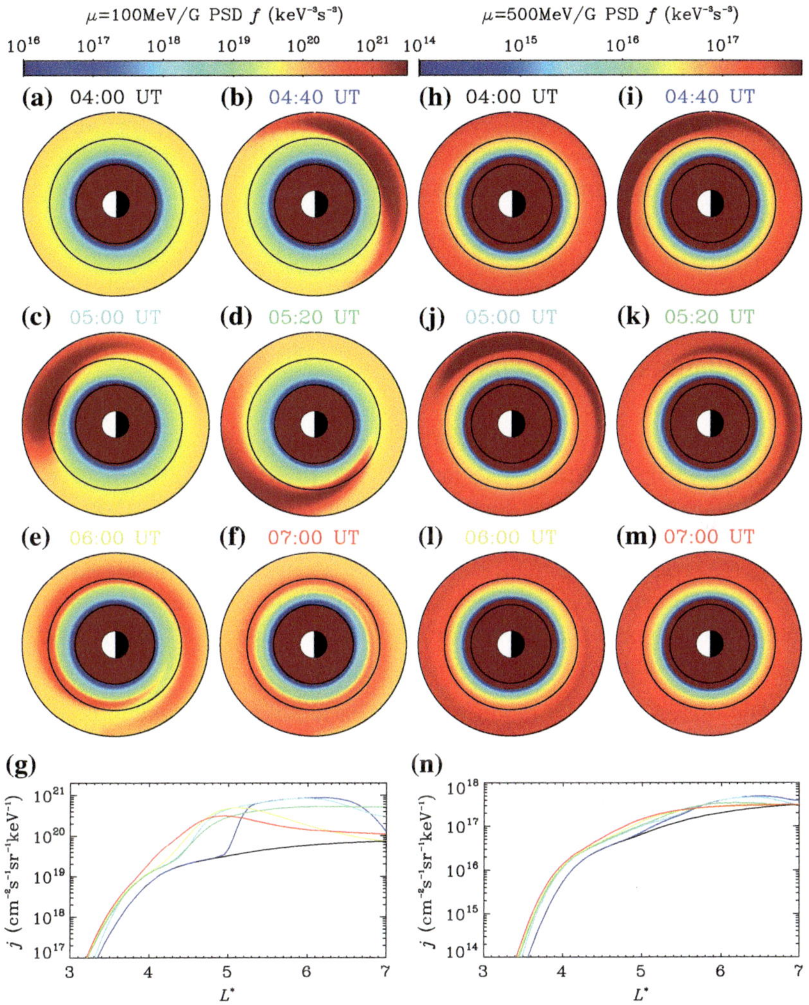

Fig. 5.4 Evolution of equatorially trapped electron PSD at $\mu = 100\,\text{MeV/G}$ (**a**–**f**) and 500 MeV/G (**h**–**m**) in the (L^*, ϕ) space, and the corresponding azimuthally averaged radial PSD profiles at $\mu = 100\,\text{MeV/G}$ (**g**) and 500 MeV/G (**n**). The *black circles* denote the locations $L^* = 1, 3, 5$ and 7. The magnetic local noon is on the *left*, dusk is at the *bottom*, midnight is on the *right*, and dawn is at the *top*

field, the drift path has its closest approach to earth at midnight. The dayside LANL satellite samples the electrons in the inner drift orbit, while the nightside LANL satellite samples the electrons in the outer drift orbit. Thus, the injected electrons will spend more time on the convection from outer orbit to inner one.

Figure 5.4 demonstrates the global distributions and the azimuthally averaged radial profiles of $\mu = 100$ and 500 MeV/G electron PSDs at the selected times.

The earthward convection and azimuthal drift of injected electrons can be clearly identified (similar to those in Fig. 5.2). Interestingly, the monotonic electron PSD at $\mu = 100\,\mathrm{MeV/G}$ becomes peaked after storm injection. A lot of ~50 keV electrons are injected from the outer boundary (see Fig. 5.1c), convected earthward, energized to ~100 keV and trapped around $L^* = 5.0$. Such process produces the peak of PSD at $\mu = 100\,\mathrm{MeV/G}$. In contrast, there is no persistently peaked PSD at $\mu = 500\,\mathrm{MeV/G}$ due to the very limited injection of >250 keV electrons at the outer boundary (see Fig. 5.2c).

5.4 Conclusions and Discussions

The kinetic radiation belt model STEERB is improved based on the bounce-averaged convection-diffusion equation. The improved STEERB model includes the adiabatic transport, magnetospheric convection, radial diffusion, and local diffusion. The adopted electromagnetic field model consists of the Hilmer-Voigt symmetric magnetic field, the typical background electric fields, and the substorm-related impulsive electric field. The radiation belt electron dynamics driven by substorm injection on 10 January 1997 is investigated via the convection-diffusion STEERB model. The main results are summarized as follows:

1. The data-driven STEERB model can more or less reproduce the injection signatures (enhancement level, drift echo, and dispersion feature) observed by three different LANL satellites at the geostationary orbit. The current kinetic simulation cannot fully answer the question of where the injected electrons at the outer boundary come from. The previous test particle simulations based on different numerical models gave different answers on this issue. Li et al. (1998) stated that 90 % of the injected electrons come from the region $\geq 9R_E$ away from the Earth, while Zaharia et al. (2004) concluded that the injected electrons have the initial positions closer than $9R_E$ from the Earth.
2. The previous observations (e.g., Miyoshi et al. 2003) have shown that the accumulation of seed electrons is important for the storm-time buildup of relativistic electron fluxes. The current substorm injection is found to be able to yield about 10 and 2 times enhancement of the azimuthally averaged 0.1 and 0.5 MeV electron fluxes within tens of minutes at the center of outer radiation belt, indicating that the substorm injection can play a significant role in the radiation belt electron dynamics by efficiently providing plenty of seed electrons for the local energization process. These results are different from those obtained on the basis of idealized simulations (Fok et al. 2001). In the previous simulations (Fok et al. 2001), the maximum enhancement of electron fluxes was located around geostationary orbit, and relatively few electrons were injected into the center of the outer radiation belt. Their simulations produced ~100 times enhancement of electron fluxes at energies up to 4 MeV due to the hard injection energy spectrum at the outer boundary.

3. The peaked electron PSDs are frequently observed in the outer radiation belt (e.g., Green and Kivelson 2004), which are often attributed to three possible physical processes: the gyro-resonant interaction with chorus waves (e.g., Summers et al. 1998; Horne et al. 2005; Shprits et al. 2006b), the losses at larger L^* (e.g., Brautigam and Albert 2000; Shprits et al. 2006a), and the drift-resonant interaction with large amplitude narrowband ULF waves (Degeling et al. 2008). Here, we identify a new candidate mechanism, substorm injection, for the generation of peaked electron PSD around $\mu = 100$ MeV/G. The freshly injected ~50 keV electrons at the outer boundary are convected earthward (primarily due to substorm-related impulsive electric field), and energized to ~100 keV at $L^* = 5$. These energized electrons are trapped with the dominance of curvature drift and grad-B drift, producing the peaked PSD in the outer radiation belt.

References

Birn J, Thomsen MF, Borovsky JE, Reeves GD, McComas DJ, Belian RD, Hesse M (1998) Substorm electron injections: Geosynchronous observations and test particle simulations. J Geophys Res 103:9235–9248. doi:10.1029/97JA02635

Bourdarie S, Boscher D, Beutier T, Sauvaud J, Blanc M (1997) Electron and proton radiation belt dynamic simulations during storm periods: a new asymmetric convection-diffusion model. J Geophys Res 102:17541–17552. doi:10.1029/97JA01305

Brautigam DH, Albert JM (2000) Radial diffusion analysis of outer radiation belt electrons during the October 9, 1990, magnetic storm. J Geophys Res 105:291–310. doi:10.1029/1999JA900344

Degeling AW, Ozeke LG, Rankin R, Mann IR, Kabin K (2008) Drift resonant generation of peaked relativistic electron distributions by Pc 5 ULF waves. J Geophys Res 113(A02):208. doi:10.1029/2007JA012411

Fok M-C, Moore TE (1997) Ring current modeling in a realistic magnetic field configuration. Geophys Res Lett 24:1775–1778. doi:10.1029/97GL01255

Fok M-C, Moore TE, Kozyra JU, Ho GC, Hamilton DC (1995) Three-dimensional ring current decay model. J Geophys Res 100:9619–9632. doi:10.1029/94JA03029

Fok M-C, Moore TE, Spjeldvik WN (2001) Rapid enhancement of radiation belt electron fluxes due to substorm dipolarization of the geomagnetic field. J Geophys Res 106:3873–3882. doi:10.1029/2000JA000150

Fok, M-C, Horne RB, Meredith NP, Glauert SA (2008) Radiation belt environment model: application to space weather nowcasting. J Geophys Res 113(A12):A03S08. doi:10.1029/2007JA012558

Fok M-C, Buzulukova N, Chen S-H, Valek PW, Goldstein J, McComas DJ (2010) Simulation and TWINS observations of the 22 July 2009 storm. J Geophys Res 115(A12):231. doi:10.1029/2010JA015443

Fok M-C, Moore TE, Slinker SP, Fedder JA, Delcourt DC, Nosé M, Chen S-H (2011) Modeling the superstorm in November 2003. J Geophys Res 116:A00J17. doi:10.1029/2010JA015720

Friedel RHW, Korth A, Kremser G (1996) Substorm onsets observed by CRRES: determination of energetic particle source regions. J Geophys Res 101:13137–13154. doi:10.1029/96JA00399

Green JC, Kivelson MG (2004) Relativistic electrons in the outer radiation belt: differentiating between acceleration mechanisms. J Geophys Res 109(A03):213. doi:10.1029/2003JA010153

Hilmer RV, Voigt G (1995) A magnetospheric magnetic field model with flexible current systems driven by independent physical parameters. J Geophys Res 100:5613–5626. doi:10.1029/94JA03139

Horne RB, Thorne RM, Shprits YY, Meredith NP, Glauert SA, Smith AJ, Kanekal SG, Baker DN, Engebretson MJ, Posch JL, Spasojevic M, Inan US, Pickett JS, Decreau PME (2005) Wave acceleration of electrons in the Van Allen radiation belts. Nature 437:227–230. doi:10.1038/nature03939

Jordanova VK, Miyoshi Y (2005) Relativistic model of ring current and radiation belt ions and electrons: initial results. Geophys Res Lett 32(L14):104. doi:10.1029/2005GL023020

Jordanova VK, Kistler LM, Kozyra JU, Khazanov GV, Nagy AF (1996) Collisional losses of ring current ions. J Geophys Res 101:111–126. doi:10.1029/95JA02000

Jordanova VK, Kozyra JU, Nagy AF, Khazanov GV (1997) Kinetic model of the ring current-atmosphere interactions. J Geophys Res 102:14279–14292. doi:10.1029/96JA03699

Jordanova VK, Boonsiriseth A, Thorne RM, Dotan Y (2003) Ring current asymmetry from global simulations using a high-resolution electric field model. J Geophys Res 108:1443. doi:10.1029/2003JA009993

Jordanova VK, Albert J, Miyoshi Y (2008) Relativistic electron precipitation by EMIC waves from self-consistent global simulations. J Geophys Res 113:A00A10. doi:10.1029/2008JA013239

Jordanova VK, Zaharia S, Welling DT (2010) Comparative study of ring current development using empirical, dipolar, and self-consistent magnetic field simulations. J Geophys Res 115:A00J11. doi:10.1029/2010JA015671

Li X, Baker DN, Temerin M, Reeves GD, Belian RD (1998) Simulation of dispersionless injections and drift echoes of energetic electrons associated with substorms. Geophys Res Lett 25: 3763–3766. doi:10.1029/1998GL900001

Maynard NC, Chen AJ (1975) Isolated cold plasma regions—observations and their relation to possible production mechanisms. J Geophys Res 80:1009–1013. doi:10.1029/JA080i007p01009

McIlwain CE (1974) Substorm injection boundaries. In: McCormac BM (ed) Magnetospheric physics, pp 143–154

Miyoshi Y, Morioka A, Misawa H, Obara T, Nagai T, Kasahara Y (2003) Rebuilding process of the outer radiation belt during the 3 November 1993 magnetic storm: NOAA and Exos-D observations. J Geophys Res 108:1004. doi:10.1029/2001JA007542

Miyoshi YS, Jordanova VK, Morioka A, Thomsen MF, Reeves GD, Evans DS, Green JC (2006) Observations and modeling of energetic electron dynamics during the October 2001 storm. J Geophys Res 111:A11S02. doi:10.1029/2005JA011351

Reeves GD, Henderson MG, McLachlan PS, Belian RD, Friedel RHW, Korth A (1996) Radial propagation of substorm injections. In: Rolfe EJ, Kaldeich B (ed) International conference on substorms, ESA special publication, vol 389, pp 579–584

Roederer JG (1970) Dynamics of geomagnetically trapped radiation. Springer, New York

Sarris TE, Li X, Tsaggas N, Paschalidis N (2002) Modeling energetic particle injections in dynamic pulse fields with varying propagation speeds. J Geophys Res 107:1033. doi:10.1029/2001JA900166

Schulz M, Lanzerotti LJ (1974) Particle diffusion in the radiation belts, in physics and chemistry in space, vol 7. Springer, New York

Shprits YY, Thorne RM, Friedel R, Reeves GD, Fennell J, Baker DN, Kanekal SG (2006a) Outward radial diffusion driven by losses at magnetopause. J Geophys Res 111(A11):214. doi:10.1029/2006JA011657

Shprits YY, Thorne RM, Horne RB, Glauert SA, Cartwright M, Russell CT, Baker DN, Kanekal SG (2006b) Acceleration mechanism responsible for the formation of the new radiation belt during the (2003) halloween solar storm. Geophys Res Lett 33(L05):104. doi:10.1029/2005GL024256

Shu C-W, Osher S (1989) Efficient implementation of essentially non-oscillatory shock-capturing schemes II. J Comput Phys 83:32. doi:10.1016/0021-9991(89)90222-2

Su Z, Xiao F, Zheng H, Wang S (2010a) STEERB: a three-dimensional code for storm-time evolution of electron radiation belt. J Geophys Res 115(A09):208. doi:10.1029/2009JA015210

Su Z, Xiao F, Zheng H, Wang S (2010b) Combined radial diffusion and adiabatic transport of radiation belt electrons with arbitrary pitch-angles. J Geophys Res 115(A10):249. doi:10.1029/2010JA015903

Su Z, Xiao F, Zheng H, Wang S (2011a) CRRES observation and STEERB simulation of the 9 (October 1990) electron radiation belt dropout event. Geophys Res Lett 38(L06):106. doi:10.1029/2011GL046873

Su Z, Xiao F, Zheng H, Wang S (2011b) Radiation belt electron dynamics driven by adiabatic transport, radial diffusion, and wave-particle interactions. J Geophys Res 116(A04):205. doi:10.1029/2010JA016228

Summers D, Thorne RM, Xiao F (1998) Relativistic theory of wave-particle resonant diffusion with application to electron acceleration in the magnetosphere. J Geophys Res 103(20):487

Thorne RM, O'Brien TP, Shprits YY, Summers D, Horne RB (2005) Timescale for MeV electron microburst loss during geomagnetic storms. J Geophys Res 110(A09):202. doi:10.1029/2004JA010882

Volland H (1973) A semiempirical model of large-scale magnetospheric electric fields. J Geophys Res 78:171–180. doi:10.1029/JA078i001p00171

Zaharia S, Birn J, Friedel RHW, Reeves GD, Thomsen MF, Cheng CZ (2004) Substorm injection modeling with nondipolar, time-dependent background field. J Geophys Res 109(A10):211. doi:10.1029/2004JA010464

Chapter 6
Summary

The electron radiation belt comprises electrons with energies from $\sim$0.1 MeV to several MeV trapped in the geomagnetic field. According to the spatial region, the electron radiation belt can be divided into inner and outer belts separated by the slot region with low level of energetic electron fluxes. The inner belt is relatively stable, but the outer belt often exhibits significant variabilities during geomagnetically active periods. The radiation belt was discovered in 1958, and has received a resurgence of interest in recent years (Friedel et al. 2002; Thorne 2010). There are two main drivers: first, the resolving of fundamental science problems surrounding the complex and dramatic dynamics of radiation belt, second, the need to reduce or even avoid the hazards of radiation belt posed to space-borne systems. Both objectives can be facilitated by the development of radiation belt physical models which has attracted the increasing attention of many research groups (e.g., Li et al. 1993; Beutier et al. 1995; Bourdarie et al. 1997; Zheng et al. 2003; Jordanova and Miyoshi 2005; Ukhorskiy et al. 2006; Fok et al. 2008; Subbotin and Shprits 2009; Subbotin et al. 2010; Su et al. 2010a, 2011b; Xiao et al. 2010b; Saito et al. 2010; Zheng et al. 2011).

6.1 Developing of STTERB Model and Obtained Physical Results

We here briefly summarize the building steps of the STEERB model and the obtained physical results at each step.

First, we develop a radiation belt local diffusion model treating the cyclotron resonance with various plasma waves. Our proposed hybrid finite difference method can overcome the potential numerical instability associated with the cross diffusion terms. This numerical scheme is efficient, stable, and easily-programmed, and has been used in many quasilinear simulations by our research group (Su 2009c; Su et al. 2009a, b, 2010a, b, c, 2011a, b, c; Xiao et al. 2009, 2010b, a, 2011) and other

Z. Su, *A Global Kinetic Model for Electron Radiation Belt Formation and Evolution*, Springer Theses, DOI 10.1007/978-3-662-46651-3_6

groups (Fok et al. 2010; Thorne et al. 2010; Tao et al. 2011; Zheng et al. 2011). The contributions of various waves to the electron radiation belt evolution are quantified, and the obtained results generally support the previous findings. The importance of cross diffusion is further evaluated by comparison, and it is found that cross diffusion plays different roles in the different wave-particle interactions.

Then, we build a global electron radiation belt diffusion model STEERB, including radial diffusion and local diffusion by cyclotron resonance. STEERB model is one of the first kinetic models to include cross diffusion. It is efficient, stable, and easily-parallelizable and reproduces the general characteristics of electron radiation belt during geomagnetically quiet and active times. The contributions of radial diffusion and various cyclotron resonance to the radiation belt evolution are identified, and particularly the importance of cross diffusion in the global diffusion simulations are evaluated.

Next, we adopt a Dst-dependent geomagnetic field, and self-consistently introduce the adiabatic transport into the STEERB model. The STEERB model is the first radiation belt model to accurately include adiabatic transport. The effect of adiabatic transport on the radiation belt electron evolution is quantified under conditions with and without other non-adiabatic processes. The electron radiation belt dropout event on 9 October 1990 is reproduced by the data-driven STEERB model, and the dominant loss mechanisms are determined in this specific event.

Finally, we improve the STEERB model into a global electron radiation belt convection-diffusion model including adiabatic transport, magnetospheric convection, radial diffusion, and cyclotron resonance. The current STEERB model is efficient, stable, and easily parallelizable, possesses high temporal and spatial resolutions, and particularly has the capability of handling transient processes. The substorm injection event on 10 January 1997 is reproduced by the data-driven STEERB model, and the contribution of substorm injection to electron radiation belt dynamics is identified.

6.2 Comparison of Radiation Belt Kinetic Models

The test-particle and kinetic formulations are two commonly used approaches for radiation belt model. Here we focus on the electron radiation belt kinetic models. These models can be classified into two types: diffusion model and convection-diffusion model. The first global diffusion model Salammbô was developed by Beutier and Boscher (1995), and further improved by Varotsou et al. (2005, 2008). More diffusion models, e.g., ALBERT (Albert et al. 2009), VERB (Shprits et al. 2009; Subbotin and Shprits 2009; Subbotin et al. 2010) and the early STEERB (Su et al. 2010a, 2011b; Xiao et al. 2010b). Afterward, some ring current models, e.g., RAM (Jordanova and Miyoshi 2005) and RBE (Fok et al. 2008) has been generalized into the convection-diffusion radiation belt models. Recently, the STEERB model has been improved into the convection-diffusion model. A comparison of these models is listed in Table 6.1. Only RBE and STEERB models adopt the time-dependent

Table 6.1 Comparison of seven radiation belt kinetic models (in chronological order)

			Physical processes[a]					
Name	Background magnetic field	Background electric field	Adiabatic	Non-adiabatic				
			AT	CR		RD	MC	
				CW	PW		SC	IC
Salammbô	Dipole	–	N	Y	Y	Y	N	N
BOURDARIE	Eccentric dipole	Volland (1973), Maynard and Chen (1975) convection and co-rotation electric fields	N	N	Y	N	Y	N
RAM	Dipole	Volland–Maynard–Chen convection and co-rotation electric fields	N	N	Y	Y	Y	N
RBE	TS05 (Tsyganenko and Sitnov 2005)	Weimer (2001) and induced electric field	N	Y	N	N	Y	Y
ALBERT	Dipole	–	N	Y	N	Y	N	N
VERB	Dipole	–	N	Y	Y	Y	N	N
STEERB	Symmetric Hilmer-Viogt (Hilmer and Voigt 1995)	Volland-Maynard-Chen convection and co-rotation electric fields, as well as the impulsive electric field associated with the substorm injection (Sarris et al. 2002)	Y	Y	Y	Y	Y	Y

[a] Involved processes include the adiabatic transport (*AT*), cyclotron resonance (*CR*), radial diffusion (*RD*) and magnetospheric convection (*MC*). The cyclotron resonance is driven by the chorus waves (*CW*) out of plasmasphere and the hiss and EMIC waves in the plasmasphere or plume (*PW*). The magnetospheric convection contains the steady convection (*SC*) related to the magnetic (gradient and curvature) drift and the traditional (convection and co-rotation) electric drift, and the impulsive convection (*IC*) associated with the substorm injection

background geomagnetic field, while other models use the fixed background geomagnetic field. Diffusion models do not need background electric fields. The RBE model adopts the most realistic electric field, other convection-diffusion models use some relatively simple electric fields. STEERB model covers the most comprehensive physical processes, and other models ignore one or more physical processes. RBE model does not explicitly include radial diffusion, but may implicitly introduce radial diffusion in the presence of appropriate electromagnetic disturbances (Fok et al. 2010). Although RBE adopts the time-dependent geomagnetic field, it does not involve the third adiabatic invariant and then does not accurately include adiabatic transport.

6.3 Future Developments of STEERB Model

There remain important improvements for the STEERB model:

1. More realistic auxiliary background models should be taken into account.
2. More observational constraints should be imposed on the inputs of STEERB model. The successful launch and stable operation of Van Allen Probes would largely promote the development of radiation belt kinetic models.
3. More physical processes, e.g., convective type of drift resonance with ULF waves (Degeling et al. 2008) and nonlinear resonance with large amplitude plasma waves (e.g., Cattell et al. 2008; Cully et al. 2008; Bortnik et al. 2008; Albert and Bortnik 2009; Yoon 2011), need to be introduced into the STEERB model.

References

Albert JM, Bortnik J (2009) Nonlinear interaction of radiation belt electrons with electromagnetic ion cyclotron waves. Geophys Res Lett 36(L12):110. doi:10.1029/2009GL038904

Albert JM, Meredith NP, Horne RB (2009) Three-dimensional diffusion simulation of outer radiation belt electrons during the october 9, 1990, magnetic storm. J Geophys Res 114(A09):214. doi: 10.1029/2009JA014336

Beutier T, Boscher D (1995) A three-dimensional analysis of the electron radiation belt by the salammbô code. J Geophys Res 100:14853–14862. doi:10.1029/94JA03066

Beutier T, Boscher D, France M (1995) Salammbo: a three-dimensional simulation of the proton radiation belt. J Geophys Res 100:17181–17188. doi:10.1029/94JA02728

Bortnik J, Thorne RM, Inan US (2008) Nonlinear interaction of energetic electrons with large amplitude chorus. Geophys Res Lett 35(L21):102. doi:10.1029/2008GL035500

Bourdarie S, Boscher D, Beutier T, Sauvaud J, Blanc M (1997) Electron and proton radiation belt dynamic simulations during storm periods: a new asymmetric convection-diffusion model. J Geophys Res 102:17541–17552. doi:10.1029/97JA01305

Cattell C, Wygant JR, Goetz K, Kersten K, Kellogg PJ, von Rosenvinge T, Bale SD, Roth I, Temerin M, Hudson MK, Mewaldt RA, Wiedenbeck M, Maksimovic M, Ergun R, Acuna M, Russell CT (2008) Discovery of very large amplitude whistler-mode waves in Earth's radiation belts. Geophys Res Lett 35(L01):105. doi:10.1029/2007GL032009

Cully CM, Bonnell JW, Ergun RE (2008) THEMIS observations of long-lived regions of large-amplitude whistler waves in the inner magnetosphere. Geophys Res Lett 35:L17S16. doi: 10.1029/2008GL033643

Degeling AW, Ozeke LG, Rankin R, Mann IR, Kabin K (2008) Drift resonant generation of peaked relativistic electron distributions by Pc 5 ULF waves. J Geophys Res 113(A02):208. doi:10.1029/2007JA012411

Fok M-C, Horne RB, Meredith NP, Glauert SA (2008) Radiation belt environment model: application to space weather nowcasting. J Geophys Res 113(A12):A03S08. doi:10.1029/2007JA012558

Fok MC, Glocer A, Zheng Q, Horne RB, Meredith NP, Albert JM, Nagai T (2010) Recent developments in the radiation belt environment model. J Atmos Sol-Terr Phys 73:1435–1443. doi: 10.1016/j.jastp.2010.09.033

Friedel RHW, Reeves GD, Obara T (2002) Relativistic electron dynamics in the inner magnetosphere—a review. J Atmos Sol-Terr Phys 64:265–282. doi:10.1016/S1364-6826(01)00088-8

Hilmer RV, Voigt G (1995) A magnetospheric magnetic field model with flexible current systems driven by independent physical parameters. J Geophys Res 100:5613–5626. doi:10.1029/94JA03139

Jordanova VK, Miyoshi Y (2005) Relativistic model of ring current and radiation belt ions and electrons: initial results. Geophys Res Lett 32(L14):104. doi:10.1029/2005GL023020

Li X, Roth I, Temerin M, Wygant JR, Hudson MK, Blake JB (1993) Simulation of the prompt energization and transport of radiation belt particles during the March 24, 1991 SSC. Geophys Res Lett 20:2423–2426. doi:10.1029/93GL02701

Maynard NC, Chen AJ (1975) Isolated cold plasma regions—observations and their relation to possible production mechanisms. J Geophys Res 80:1009–1013. doi:10.1029/JA080i007p01009

Saito S, Miyoshi Y, Seki K (2010) A split in the outer radiation belt by magnetopause shadowing: test particle simulations. J Geophys Res 115(A08):210. doi:10.1029/2009JA014738

Sarris TE, Li X, Tsaggas N, Paschalidis N (2002) Modeling energetic particle injections in dynamic pulse fields with varying propagation speeds. J Geophys Res 107:1033. doi:10.1029/2001JA900166

Shprits YY, Subbotin D, Ni B (2009) Evolution of electron fluxes in the outer radiation belt computed with the VERB code. J Geophys Res 114(A11):209. doi:10.1029/2008JA013784

Su Z, Zheng H, Wang S (2009a) Dynamic evolution of energetic outer zone electrons due to whistler-mode chorus based on a realistic density model. J Geophys Res 114(A07):201. doi: 10.1029/2008JA014013

Su Z, Zheng H, Wang S (2009b) Evolution of electron pitch angle distribution due to interactions with whistler-mode chorus following substorm injections. J Geophys Res 114(A08):202. doi: 10.1029/2009JA014269

Su Z, Zheng H, Xiong M (2009c) Dynamic evolution of outer radiation belt electrons due to whistler-mode chorus. Chinese Phys Lett 26(039):401

Su Z, Xiao F, Zheng H, Wang S (2010a) STEERB: a three-dimensional code for storm-time evolution of electron radiation belt. J Geophys Res 115(A09):208. doi:10.1029/2009JA015210

Su Z, Zheng H, Wang S (2010b) A parametric study on the diffuse auroral precipitation by resonant interaction with whistler-mode chorus. J Geophys Res 115(A05):219. doi:10.1029/2009JA014759

Su Z, Zheng H, Wang S (2010c) Three dimensional simulations of energetic outer zone electron dynamics due to wave-particle interaction and azimuthal advection. J Geophys Res 115(A06):203. doi:10.1029/2009JA014980

Su Z, Xiao F, Zheng H, Wang S (2011a) CRRES observation and STEERB simulation of the 9 (October 1990) electron radiation belt dropout event. Geophys Res Lett 38(L06):106. doi: 10.1029/2011GL046873

Su Z, Xiao F, Zheng H, Wang S (2011b) Radiation belt electron dynamics driven by adiabatic transport, radial diffusion, and wave-particle interactions. J Geophys Res 116(A04):205. doi: 10.1029/2010JA016228

Su Z, Zheng H, Chen L, Wang S (2011c) Numerical simulations of storm-time outer radiation belt dynamics by wave-particle interactions including cross diffusion. J Atmos Sol-Terr Phys 73:95–105. doi:10.1016/j.jastp.2009.08.002

Subbotin D, Shprits Y, Ni B (2010) Three-dimensional VERB radiation belt simulations including mixed diffusion. J Geophys Res 115(A03):205. doi:10.1029/2009JA015070

Subbotin DA, Shprits YY (2009) Three-dimensional modeling of the radiation belts using the versatile electron radiation belt (VERB) code. Space Weather 7(S10):001. doi:10.1029/2008SW000452

Tao X, Thorne RM, Li W, Ni B, Meredith NP, Horne RB (2011) Evolution of electron pitch angle distributions following injection from the plasma sheet. J Geophys Res 116(A4):A04,229

Thorne RM (2010) Radiation belt dynamics: the importance of wave-particle interactions. Geophys Res Lett 37(L22):107. doi:10.1029/2010GL044990

Thorne RM, Ni B, Tao X, Horne RB, Meredith NP (2010) Scattering by chorus waves as the dominant cause of diffuse auroral precipitation. Nature 467:943. doi:10.1038/nature09467

Tsyganenko NA, Sitnov MI (2005) Modeling the dynamics of the inner magnetosphere during strong geomagnetic storms. J Geophys Res 110(A03):208. doi:10.1029/2004JA010798

Ukhorskiy AY, Anderson BJ, Brandt PC, Tsyganenko NA (2006) Storm time evolution of the outer radiation belt: transport and losses. J Geophys Res 111:A11S03. doi:10.1029/2006JA011690

Varotsou A, Boscher D, Bourdarie S, Horne RB, Glauert SA, Meredith NP (2005) Simulation of the outer radiation belt electrons near geosynchronous orbit including both radial diffusion and resonant interaction with whistler-mode chorus waves. Geophys Res Lett 32(L19):106. doi:10.1029/2005GL023282

Varotsou A, Boscher D, Bourdarie S, Horne RB, Meredith NP, Glauert SA, Friedel RH (2008) Three-dimensional test simulations of the outer radiation belt electron dynamics including electron-chorus resonant interactions. J Geophys Res 113(A12):212. doi:10.1029/2007JA012862

Volland H (1973) A semiempirical model of large-scale magnetospheric electric fields. J Geophys Res 78:171–180. doi:10.1029/JA078i001p00171

Weimer DR (2001) An improved model of ionospheric electric potentials including substorm perturbations and application to the geospace environment modeling November 24, 1996, event. J Geophys Res 106:407C416. doi:10.1029/2000JA000604

Xiao F, Su Z, Zheng H, Wang S (2009) Modeling of outer radiation belt electrons by multidimensional diffusion process. J Geophys Res 114(A03):201. doi:10.1029/2008JA013580

Xiao F, Chen L, Su Z, Zheng H, Wang S (2010a) A parametric study on outer radiation belt electron evolution by superluminous R-X mode waves. J Geophys Res 115(A10):217. doi:10.1029/2010JA015374

Xiao F, Su Z, Zheng H, Wang S (2010b) Three-dimensional simulations of outer radiation belt electron dynamics including cross diffusion terms. J Geophys Res 115(A05):216. doi:10.1029/2009JA014541

Xiao F, Chen L, He Y, Su Z, Zheng H (2011) Modeling of precipitation loss of ring current protons by electromagnetic ion cyclotron waves. J Atmos Sol-Terr Phys 73:106–111. doi:10.1016/j.jastp.2010.01.007

Yoon PH (2011) Large-amplitude whistler waves and electron acceleration. Geophys Res Lett 381(L12):105. doi:10.1029/2011GL047893

Zheng Q, Fok M, Albert J, Horne RB, Meredith NP (2011) Effects of energy and pitch angle mixed diffusion on radiation belt electrons. J Atmos Sol-Terr Phys 73:785–795. doi:10.1016/j.jastp.2011.01.014

Zheng Y, Fok M-C, Khazanov GV (2003) A radiation belt-ring current forecasting model. Space Weather 1:1013. doi:10.1029/2003SW000007